Contaminated Land

CONTAMINATED LAND

Managing Legal Liabilities
Second Edition

Edited by
Martin Warren

EARTHSCAN
Earthscan Publications Ltd, London and Sterling, VA

EVERSHEDS
Business Lawyers in Europe

Second edition, first published in the UK and USA in 2001 by
Earthscan Publications Ltd

A catalogue record for this book is available from the British Library

ISBN: 1 85383 747 4

Printed and bound by Selwood Printing Ltd, Burgess Hill, West Sussex
Cover design by Andrew Corbett

For a full list of publications please contact:

Earthscan Publications Ltd
120 Pentonville Road
London, N1 9JN, UK
Tel: +44 (0)20 7278 0433
Fax: +44 (0)20 7278 1142
Email: earthinfo@earthscan.co.uk
http://www.earthscan.co.uk

22883 Quicksilver Drive, Sterling, VA 20166–2012, USA

Earthscan is an editorially independent subsidiary of Kogan Page Ltd and publishes in association with WWF-UK and the International Institute for Environment and Development

This book is printed on elemental chlorine-free paper

TABLE OF CONTENTS

LIST OF FIGURES

PREFACE

This book is not intended to be a critique of the Government's attempt to produce a regime which will force the remediation of land spoiled by our industrial past, nor is it simply a reproduction of the Act and its Guidance. It is intended to be a guide to the new regime linking the Guidance with the relevant Sections of the Act and outlining the practical effect of the regime on those affected by it.

The contaminated land regime has been effective in England for a year, in Scotland for a little over eight months and, it is said, will be in place in Wales from July 2001. In that time we have seen little activity save for the actions of local authorities striving to develop and publish their strategies for the identification of contaminated land within their areas before the deadline of 30 June 2001.

The creation of these strategies has not been a simple task despite the provision of a document published jointly by the DETR and the Environment Agency, entitled *Contaminated Land Inspection Strategies: Technical Advice for Local Authorities*, in April 2001. Indeed, a number of large local authorities have yet to have even a final draft of their strategy approved by their Council. If this is because they are giving the issue a great deal of thought and consideration, that is heartening. If it is because of the lack of resources being made available to the project, that does not bode well for the future effectiveness of the regime.

In May 2000 the Environment Agency published its Policy in respect of the regime, which provides that the regime will be applied in a manner which, inter alia, encourages voluntary action by the polluters, reduces or removes perceived blight associated with land contamination, and seeks to place the economic burden of dealing with historical pollution on the polluter, where practicable.

This Policy also refers to the Agency/Local Government Association Protocol on Land Contamination contained in the Memorandum of Understanding between those parties and the Procedural Notes that the Agency has prepared on such topics as 'Local Authority Inspection Strategies' and 'Identification of Contaminated Land and Special Sites'.

As is clear, the regime has, to date, been the subject of much discussion and comment, but has yet to have been as active as a number of commentators expected. Whether this will be the case in the coming years we shall to wait to see.

April 2001

CHAPTER 1
CONTAMINATED LAND: AN INTRODUCTION TO THE PROBLEM

A common sense, that is non-statutory, definition of contaminated land is any land that contains a substance or substances that may adversely affect people or the environment.

With its industrial history, it is evident that Britain's "green and pleasant land" has been to a greater extent and is continuing to be to a lesser extent contaminated. The major environmental issue of recent years relates to the extent of such contamination and the question of who should pay for it's clean-up.

Environmental groups have sought to put pressure on the Government and industry to remediate such land whereas industry has adopted a pragmatic approach arguing that the legacy of the industrial revolution cannot be cleaned up in the short term for both practical and financial reasons.

Initial Approach

The Government in its White Paper of 1 September 1990 looked at the issue of contaminated land in the context of a wide-ranging review of environmental issues affecting the UK and commented generally in relation to UK environmental policy:

> *"Action on the environment has to be proportionate to the costs involved and to the ability of those expected to pay them. So it is particularly important for Government to adopt the most cost-effective instruments for controlling pollution and tackling environmental problems. And we need to ensure we have a sensible order of priorities, acting first to tackle problems that could cause most damage to human life or health and could do most damage to the environment now or in the future."*

In relation to contaminated land and the inherent difficulties in dealing with it, it was said that:

> *"contamination of land by chemicals and waste products is hard to define and measure exactly. But surveys suggest that over half of derelict land may be contaminated and contamination is also found on other land. The nature of the contamination and possible risks to health and groundwater supplies vary widely and the Government needs better assessments of the scale of the problem."*

The "scale of the problem" had been considered by a House of Commons Environment Committee in 1989 which had before it details of surveys carried out in 1988 which showed that over 50,000 hectares of derelict land existed in England alone. The subsequent Report suggested that the principle of "caveat emptor" (let the buyer beware) should be abolished in favour of more information being made available to purchasers and the public by way of a register of land uses.

Despite it being estimated by the CBI that simply to investigate the extent of contamination of derelict land in England would cost £600 million, the Government accepted the proposal of environmental information being provided to a potential purchaser by establishing, under Section 143 of the Environmental Protection Act 1990, a duty on local authorities to compile and maintain registers of land which may be contaminated.

The proposal to establish such registers became extremely controversial as various sections of industry argued that registration would have an adverse impact on property values that would be out of all proportion to the benefit of such registers in informing potential purchasers of possible contamination.

The Retreat from Section 143

Due to the perceived difficulties of compiling and maintaining such registers and the criticism of the scheme, in March 1993 the Government announced that the scheme had been withdrawn and that a new "wide ranging review" would be undertaken.

Michael Howard, the then Secretary of State for the Environment, stated on 27 March 1993:

> *"The Government remains determined to ensure that land contamination does not give rise to unacceptable risks to health and safety, to groundwater and to the environment generally, and also to ensure that vacant urban and industrial land can be put back into good use wherever practicable."*

The results of the review manifested themselves in March 1994, when the Government issued the Paper "Paying for our Past" which contained a series of provisional conclusions as to how contaminated land should be tackled. That Paper proposed the concept of "suitable for use" (which is examined in Chapter 2), in which contaminated sites were to be tackled on a site specific basis and clean-up work undertaken where there was a cost benefit in so doing.

In November 1994 the Government announced its final proposal in the Paper entitled "Framework for Contaminated Land" which confirmed its commitment to the "suitable for use" approach. This was seen as supporting sustainable development by both reducing the damage from past activities and by permitting contaminated land to be kept in or returned to beneficial use wherever practicable, thereby minimising avoidable pressures for new developments to take place on green field sites.

The Present Position

Following hard on the heels of the "Framework for Contaminated Land" Paper, the Environment Bill (now the Environment Act 1995 and hereinafter called the "1995 Act") began its passage through the House of Lords in December 1994.

The 1995 Act, which received Royal Assent on 17 July 1995, inserted provisions into the Environmental Protection Act 1990 (hereinafter called the "1990 Act") dealing with the classification, regulation and control of contaminated land in England and Wales.

The 1990 Act thus contains the definition of contaminated land for the purposes of legal control and sets out the framework under which local authorities and the Environment Agency should together identify contaminated land and require clean-up through the service of remediation notices on "appropriate persons".

Chapter 3 looks at the intricacies of the legal definition of contaminated land, Chapter 4 then looks at the powers of the local authority and Environment Agency to serve remediation notices and the contents of such notices and Chapter 5 deals with the definition of "appropriate person" for the purposes of the legislation.

Water Pollution and Contaminated Land

The 1995 Act also contains important new provisions that add to existing provisions dealing with water pollution. Section 161 of the Water Resources Act 1991, under which in specific circumstances the Environment Agency can take action to prevent or clean-up water pollution and then charge the cost of doing so to the polluter, is strengthened. The new provisions enable the Environment Agency to serve a "works notice" requiring the polluters to undertake the specified works themselves.

There are, however, important differences between these powers and the powers of the local authority or Environment Agency to serve remediation notices under the 1990 Act, as for such notices to be served the land must meet the definition of contaminated land and have been duly classified. For a works notice to be served this is not necessary. Such a notice can be served in the event of actual or potential pollution of controlled waters, which is thus a much wider set of circumstances. Chapter 8 deals with the inter-relationship between the works notice procedure and the contaminated land regime.

Guidance and Regulations

One final point needs to be made in relation to the provisions that have been inserted into the 1990 Act. Whilst the provisions are detailed and do appear complex at first reading, they are far from complete in the sense of providing all the answers to questions which will be raised not only by the enforcing authorities but also by those upon whom remediation notices are served.

The provisions in a number of key areas expressly refer to the fact that the local authority or the Environment Agency must have regard to the Guidance issued by the Secretary of State for the Environment. There are, however, two different levels of guidance. First, there is guidance which the local authority must "act in accordance" with, in other words follow fairly closely. Secondly, however, there is guidance which the local authority or Environment Agency must "have regard to", which implies that whilst relevant the local authority or Environment Agency will on occasion be able to depart from it should it feel that there are good reasons for doing so.

The current Guidance was published in its final form in March 2000 and contains three key areas where the local authority must "act in accordance" with it. These areas are firstly in relation to determining what is "significant" in relation to the contaminated land definition and, indeed, in determining whether pollution of controlled waters is being, or is likely to be, caused under Section 78A of the 1990 Act; secondly, in relation to the identification of contaminated land under Section 78B; and thirdly in relation to the apportionment of liability and costs between appropriate persons under Section 78F.

In respect of the "have regard to" element, this has effect in two areas; firstly the remediation of contaminated land under Section 78E; and secondly in relation to recovering costs under the provisions of Section 79P.

It is important to bear these matters in mind as failure by either enforcing authority to consider the Guidance in accordance with these requirements is a specific ground for appeal against the service of a remediation notice.

In addition to the Guidance, further technical detail is contained within the Contaminated Land (England) Regulations 2000 (hereinafter called the "2000 Regulations") which came into force on 1 April 2000. These Regulations deal with the fine detail of issues such as the form and content of remediation notices (Regulation 4) and the appeal provisions (Regulations 7–14).

Criminal Liability

To support the power of the local authority and the Environment Agency, Section 78M of the 1990 Act creates a criminal offence of failing, without reasonable excuse, to comply with the requirements of a remediation notice. This means that if a remediation notice is properly served and then not complied with, it is possible for either the local authority, or the Environment Agency where appropriate, to commence a criminal prosecution which will result in the company or individual concerned being required to appear in the Magistrates' Court and face a fine of up to £20,000 (where the remediation notice relates to industrial, trade or business premises which will be the majority of cases) together with a continuing daily fine of up to £2,000 until the remediation notice is complied with or until the local authority or the Environment Agency decide to carry out the work themselves. If they do carry out the work themselves both enforcing authorities are entitled to recover their reasonable costs in so doing from the appropriate person.

For example, company "A" is a chemical manufacturer and storage company that owns an industrial site which the local authority has decided falls within the definition of contaminated land and has therefore served a remediation notice on them. The company disagrees with the local authority's opinion but takes no formal steps to appeal against the service of the notice. By burying its head in the sand the company places itself in a difficult position because it can be summoned to appear before the Magistrates' Court to face a fine and the attendant bad publicity; furthermore, the local authority can undertake the work which was specified in the remediation notice itself and look to the company to settle the costs of such remediation.

It is essential under the new regime that the persons who are affected understand their position and know what must be done to minimise their potential liabilities.

Civil Liability

It is very important to appreciate that, despite the newly effective provisions contained within the 1990 Act, there is also a regime of civil liability under the common law.

All individuals and companies have the potential to be pursued in the civil courts for damages (that is, financial compensation) where their actions or omissions have caused land owned by other persons to become contaminated.

The two most relevant arguments that will be pursued before the civil courts are negligence and nuisance. These important principles were tested in 1993 in a case which reached the House of Lords (*Cambridge Water Company* v *Eastern Counties Leather*) and the result of this case, which is the definitive position in relation to civil liability, will be considered in Chapter 6.

Chapter 9 contains practical advice in relation to the selling or buying of contaminated or potentially contaminated land.

CHAPTER 2
SUITABLE FOR USE: THE CURRENT APPROACH

After the publication of the "Framework for Contaminated Land" document in November 1994 and much subsequent dallying, the Guidance to the contaminated land provisions in the 1990 Act was produced which stated the Government's objectives with respect to contaminated land and its approach to remediation.

The Guidance states the objectives as:

> *(a) to identify and remove unacceptable risks to human health and the environment;*
> *(b) to seek to bring damaged land back into beneficial use; and*
> *(c) to seek to ensure that the cost burden faced by individuals, companies and society as a whole are proportionate, manageable and economically sustainable.*

It is further stated that these objectives underlie the "suitable for use" approach to the remediation of land adopted by the Government to achieve sustainable development and which means that a balance must be drawn between remediating past pollution and the costs of doing so.

The "suitable for use" approach combines three elements:

1. Ensuring land is suitable for its current purpose.

   That is ensuring it is not causing unacceptable risk to human health and the environment based on its current use.

2. Ensuring that land is made suitable for any new use, as official permission is given for that new use.
   For example, where development is proposed there will need to be an assessment of potential risk with a view to remediation before the new use commences.

3. Limiting requirements for remediation to the work necessary to prevent unacceptable risk to uman health or the environment in relation to the current use or officially permitted future use of the land.

   That is, levels of remediation will reflect the current use or proposed use not *any* use.
   It is believed that:

   > *".... limiting remediation costs to what is needed to avoid unacceptable risks will mean that we will be able to recycle more previously developed land than would otherwise be the case, increasing our ability to make beneficial use of the land. This helps to increase the social, economic and environmental benefits from regeneration projects and to reduce unnecessary development pressures on green field sites."*

There is one exception to this approach and that is where the contamination has resulted from a breach of an environmental licence or permit. In such cases the contamination should be removed completely as to do otherwise would "be to undermine the regulatory regime aimed at preventing new contamination".

Whilst this approach is seen as being the most practicable, there are inherent difficulties which will generate much argument and which will only be clarified by the courts or by the issuing of subsequent guidance. For example, "current use" is defined in the Guidance as meaning not only the present use but also any use which is "likely" to be made of the land, yet there is no definition of the word "likely". Is the subjective judgement to be more likely than not (a greater than 50% chance of it occurring) or is a 10% probability acceptable?

The same problem arises in relation to the likelihood of risk (which will be considered in the next chapter).

While such issues may appear to be minor, they are issues that will form the substance of future appeals quite simply because of the likely costs involved in remediating the land.

CHAPTER 3
THE IDENTIFICATION OF CONTAMINATED LAND

Local Authority Inspection

Section 78B of the 1990 Act requires local authorities to "cause" their areas to be inspected for possible contaminated sites and in so doing they must act in accordance with the Guidance. In relation to such inspections, the Guidance suggests that the local authorities should take a "strategic approach" to the identification of such land and should prepare a written strategy for publication before July 2001.

In developing this approach the strategy should take into account local circumstances including:

- the extent to which information on land contamination is already available;
- any available evidence that significant harm or pollution of controlled waters is actually being caused;
- the history, scale and nature of industrial or other activities which may have contaminated the land in different parts of its area;
- the extent to which vulnerable targets are likely to be in contact with the contaminated land.

Thus the public are provided with a document which, although not entitled a "Register of Contaminated Land", may provide to some extent, directly or indirectly, the same information as would have been contained within the discarded Section 143 registers.

Having said that there is a duty to inspect, the question of course is, "what is the local authority looking for?"

Defining Contaminated Land

Section 78A (2) of the 1990 Act defines contaminated land as:

> *"... any land which appears to the local authority in whose area it is situated to be in such a condition, by reason of substances in, on or under the land, that:*
>
> *(a) significant harm is being caused or there is a significant possibility of such harm being caused; or*
> *(b) pollution of controlled waters is being, or is likely to be, caused."*

In determining whether land meets that definition the local authority, whilst acting in accordance with the Guidance, must address three questions:

(a) What harm is to be regarded as "significant";
(b) Whether the possibility of significant harm being caused is "significant";
(c) Whether pollution of controlled waters is being or is likely to be caused.

As the Guidance itself makes clear, the definition of contaminated land is based on the principles of risk assessment, with risk being defined as the combination of the probability of an occurrence of a specified hazard and the seriousness of the consequences.

Risk Assessment

The Guidance introduces the concept of a contaminant-pathway-receptor approach. This requires the identification of a receptor (target), for example a child, then identifying the contaminant and then establishing a plausible pathway between the contaminant and the target.

As the Guidance states:

> *"The identification of each of these three elements is linked to the identification of the others. A pathway can only be identified if it is capable of exposing an identified receptor to an identified contaminant. That particular contaminant should likewise be capable of harming or, in the case of controlled waters, be capable of polluting that particular receptor."*

Within the Guidance, this identification of the contaminant-pathway-receptor stage is called the "pollutant linkage" and where no such linkage arises the land should not be identified as contaminated land for the purposes of Section 78A (2) of the 1990 Act.

Once such a pollutant linkage has been identified, the second step for the local authority is for it to satisfy itself that the linkage:

- is resulting in significant harm being caused to the receptor in the pollutant linkage;
- presents a significant possibility of significant harm being caused to that receptor;
- is resulting in the pollution of the controlled waters which constitute the receptor; or
- is likely to result in such pollution.

Is the Risk Significant?

Having established that the land appears to be contaminated the local authority then has to determine whether or not there is significant harm or whether there is potential for significant harm being caused.

Section 78A (4) defines "harm" as:

> *"harm to the health of living organisms or other interference with the ecological systems of which they form part and, in the case of man, includes harm to his property."*

In terms of significant harm the Guidance provides that harm is only significant if it falls within Table A of the Guidance which specifies four categories of receptor with attendant relevant descriptions of harm, the first being the most extreme or significant, the last being the least extreme or significant, which are broadly as follows:

- harm to human health;
- harm to any ecological system within specified categories of location or living organism forming part of such a system which leads to an irreversible adverse change;
- harm to property including domestic animals, livestock and crops;
- harm to property in the form of buildings including a substantial interference of any right of occupation.

Is the Possibility Significant?

When deciding whether the possibility or risk of significant harm being caused is significant the local authority are, in accordance with the Guidance, to consider:

- the nature and degree of the harm;
- the susceptibility of the receptor to which the harm will be caused;
- the timescale within which the harm might occur.

The Guidance goes on to identify in a second table (Table B) the conditions which are to be met for there to be a significant possibility of such harm. These are broadly:

- human health effects via ingestion of the contaminant resulting in an unacceptable intake; such risk to be assessed on the basis of the toxicological property of that pollutant;
- all other human health effects (particularly by way of explosion or fire) if the probability of occurrence of such harm is unacceptable based on the type of pollutant linkage or the type of significant harm;
- all ecological system effects if the significant harm is more likely than not to result from the pollutant linkage in question or there is a reasonable possibility of such harm being caused and if it did it would result in damage to features of special interest;
- all animal crop effects if the significant harm is more likely than not;
- all building effects if the significant harm is more likely than not during the expected economic life of the building.

In making any assessment of what is unacceptable in relation to risk to human health, it is suggested that factors the enforcing authority should consider are:

- the level of protection already afforded;
- the way in which the substance present on, in or under the land could give rise to harm, including synergistic or antagonistic effects resulting from the presence of different substances;
- the time-scale within which the harm may occur;
- whether the harm would be irreversible;
- whether a substantial number of persons would be affected.

Pollution of Controlled Waters

The definition of contaminated land also applies to actual or potential pollution of "controlled waters" as defined in the Water Resources Act 1991 (hereinafter called the "1991 Act") and which effectively includes all waters, streams, rivers, ponds, coastal waters and, significantly, groundwater.

Pollution of controlled waters is defined at Section 78A (9) of the 1990 Act as having the same meaning as in the 1991 Act and is thus the entering into controlled waters of any poisonous, noxious or polluting matter or any solid waste matter. The enforcing authority should be satisfied that the substance is continuing to enter or is likely to enter, thereby excluding substances which have already entered.

The effect of this definition is that where land falls within the statutory definition of contaminated land due to the actual or potential pollution of controlled waters, it is not clear what level of pollution of controlled waters has to result for the definition to be met. If the definition were taken at face value, *any* poisonous, noxious or polluting matter in controlled waters would trigger the definition irrelevant of the quantity concerned. However, whilst the Guidance acknowledges that it cannot influence the question of what constitutes pollution of controlled waters, it indirectly supports the proposition that not all pollution of the controlled waters would trigger the definition by referring to the fact that any required remediation must be reasonable having regard to the cost involved and the seriousness of the pollution.

Indeed, the Guidance acknowledges that the lack of guidance concerning such pollution can be problematical and refers to the indication given by the Government that the primary legislation may be amended to deal with this specific issue.

It is apparent, however, that in determining whether pollution of controlled waters exist or is likely to occur, the local authority will take into consideration the following factors:

- whether the substances are present in, on or under the land which would be regarded as, or could create, poisonous, noxious or polluting matter;
- the mobility or other relevant properties of such substances;
- the conditions at the site;
- the underlying hydrogeology;
- the proximity of surface waters;
- the proximity of ground waters;
- the sensitivity of such surface or ground waters, that is whether any relevant water quality objectives are or are likely to be exceeded.

The local authority must consult the Environment Agency to confirm whether or not the degree of pollution of controlled waters is sufficient to enable the local authority to categorise land as contaminated for that reason.

It is interesting to note that the Guidance indicates that the meaning of "likely" in respect of the pollution of controlled waters means "on the balance of probabilities", that is, more likely than not, rather than leaving the definition to be argued.

Record of Determination

The Guidance requires the local authority to prepare a written record of any determination that particular land is contaminated land for the purposes of the 1990 Act. This record should include:

- a description of the particular pollutant linkage, identifying all three components of pollutant, pathway and receptor;
- a summary of the evidence upon which the determination is based;
- a summary of the relevant assessment of this evidence;
- a summary of the way in which the authority considers that the requirements of the Guidance have been satisfied.

It is clear from this that local authorities will be required to undertake a sophisticated risk assessment exercise before deciding to designate any site as contaminated land and be in a position to justify with evidence their thinking and decision-making in each stage of the process.

This record will be available to public scrutiny by virtue of the Environmental Information Regulations 1992, which require subject to specific exceptions, local authorities to disclose information held by them relating to the environment.

This means that the designation process is exposed to an application for judicial review in the High Court where it is believed that the decision making process has not been undertaken properly. Such an application for judicial review will, in appropriate cases, mean that no further enforcement action will be able to be taken in relation to the alleged contaminated land until such time as the High Court has determined the issue.

Local authorities are used to the threat of such proceedings and it remains to be seen whether the local authorities efforts to designate sites as contaminated land sites will be hindered by such applications.

The greatest difficulty facing enforcing authorities however is the fact that they cannot recover the costs of classifying land as contaminated. They are only entitled to recover the costs of remediation which is likely to lead to the regime being reactive rather than proactive.

CHAPTER 4
THE CLEAN-UP MECHANISM: REMEDIATION NOTICES

Defining Remediation

The remediation notice is the method by which the enforcing authority requires an "appropriate person" to undertake work in relation to a contaminated site. Failure to comply without due cause is a criminal offence which as previously stated may lead to a fine of £20,000 on conviction and a continuing fine of up to 10% of that sum for each day of non-compliance thereafter. Indeed, in extreme cases the enforcing authority is able to obtain a High Court Order requiring compliance, breach of which would be contempt of court and punishable with imprisonment.

The 1990 Act defines remediation widely (Section 78A (7)):

> *(a) the doing of anything for the purpose of assessing the condition of*
>
> *(i) the contaminated land in question;*
> *(ii) any controlled waters affected by that land; or*
> *(iii) any land adjoining or adjacent to that land;*
>
> *(b) the doing of any works, the carrying out of any operations or the taking of any steps in relation to any such land or waters for the purpose of:*
>
> *(i) preventing or minimising, or remedying or mitigating the effects of, any significant harm, or any pollution of controlled waters, by reason of which the contaminated land is such land; or*
> *(ii) of restoring the land or waters to their former state; or*
> *(iii) the making of subsequent inspections from time to time for the purpose of keeping under review the condition of the land or waters...*

Thus, remediation includes all forms of investigatory work, "pure" remediation work and also periodic inspections of the land once the work has been undertaken.

It is important to note that once land has been determined to fall within the definition of contaminated land, there is a duty on the enforcement authority to serve a remediation notice (Section 78E (1)) on the "appropriate person" (subject to the provisions of Section 78H (see Consultation below)) and where there is more than one appropriate person, different remediation notices can be served on each appropriate person requiring them to undertake relevant and perhaps different work (Section 78E (2)).

It is also clear that where two or more appropriate persons have by their actions combined to cause the contamination in question, remediation notices served on them must state the proportion of the cost which each appropriate person must bear (Section 78E (3)). If, therefore, in a situation where land is contaminated by three chemicals as a result of three different activities, it is possible (assuming that the three persons are appropriate persons) for each to be pursued by way of a remediation notice in respect of the contamination which they and they alone caused.

It is not possible for a remediation notice to specify clean-up work with no regard to the cost involved. Section 78E (4) of the 1990 Act makes it clear that remediation notices can only require work to be done which is reasonable, having regard to the likely cost and the seriousness of the harm or pollution of controlled waters. The Guidance spells out in some detail the type of work which can be specified in a remediation notice, the standards of remediation and, perhaps most importantly, what type of work is and what type of work is not to be regarded as reasonable.

What is Reasonable?

The Guidance states that:

> *"The enforcing authority should regard a remediation action as being reasonable...if an assessment of the costs likely to be involved and of the resulting benefits shows that those benefits justify incurring those costs."*

Benefits are to be regarded as contributions to the reduction of the seriousness of the harm or pollution of controlled waters or the mitigation of the seriousness of any effects of such harm or pollution. When considering the cost element of the balancing exercise, the enforcing authority should take into account:

- all the initial costs of a remediation, for example, a feasibility study;
- on-going management costs; and
- any relevant disruption costs, for example, depreciation in land values or other interests.

Standard of Remediation

The aim of remediation should be to ensure that the land in its current use is no longer "contaminated land" and that the effects of any significant harm or pollution of controlled waters that has occurred are remedied.

This can be achieved by ensuring that the "pollutant linkage" is no longer significant by either removing or treating the pollutant, breaking the pathway or removing or protecting the receptor/target *and* remedying the effect of the significant pollutant linkage using the best practicable techniques.

In deciding what are the" best practicable techniques", the enforcing authority should work on the basis of authoritative scientific and technical advice; whether that advice is obtained internally or externally will be a question for the authority.

Where even by using best practicable techniques the stated aim cannot be achieved, the standard of remediation is that which comes as close as practicable to achieving that aim *and* does remedy the effect of the significant pollutant linkage *and* puts in place a system to remedy any significant harm or pollution of controlled waters which may occur in the future due to the continued existence of the pollutant linkage. It is evident that "best practice" are the watchwords so far as remediation is concerned. The Contaminated Land (England) Regulations 2000, detail precisely the form and content of 2000 remediation notices and the associated procedural requirements, for example, the grounds of appeal against a notice and the registers which are to be kept by the enforcing authorities.

Limits on Enforcing Authorities' Powers

There are restrictions on the serving of remediation notices and these are contained in section 78H of the 1990 Act and these are:

- where it would not be reasonable to do so (see above);
- where the enforcing authority is satisfied that appropriate work is being or will be done by way of remediation without the necessity of serving a notice;
- where the appropriate person upon whom to serve a notice is the authority itself;
- where the authority considers that it should exercise it's own clean-up powers under section 78N (see Chapter 5).

Where the first of these applies the enforcing authority must issue a "remediation declaration" explaining why the authority believed it could have acted and detailing why it decided it shouldn't. For the others a "remediation statement" must be drawn-up describing the work to be done.

The "remediation statement" is prepared by the enforcing authority except where the remediation work is being undertaken by the appropriate person, in which case that person prepares it.

These documents must be entered in the public register maintained under section 78R of the 1990 Act (see Chapter 5).

Consultation

There is a further restriction on the serving of a remediation notice relating to time-scale. During the passage of the Environment Bill through Parliament there was some criticism that enforcing authorities were being given no flexibility in the service of remediation notices and that they would be required to serve such notices "immediately" upon being satisfied that the land itself fell within the statutory definition. As a result, Section 78H as inserted into the 1990 Act, contains a provision requiring the enforcing authority to consult with the person upon whom it is intended to serve the notice (the "appropriate person"), together with any owner or occupier of the land in relation to what is to be done by way of remediation, the aim of the legislation being to facilitate an informed discussion prior to the service of the formal notice.

The Guidance suggests that this consultation stage will be useful for the enforcing authority as it will allow them to revisit the question of whether the land should be identified as contaminated and also may lead to an agreed remediation programme without the necessity of serving a notice.

Following the consultation process a remediation notice may be served but not before three months have expired since the appropriate person was formally notified that the enforcing authority considered the land to be contaminated. It should not, however, be assumed that the 1990 Act is creating a strict three-month consultation period. The Act says a remediation notice may not be served within three months of notification — it does not say however that such a notice *must* be served three months *after* notification. If an appropriate person therefore feels further consultation is required they should press for it notwithstanding that technically a remediation notice could be served. However, in circumstances of imminent danger of serious harm or serious pollution of controlled waters, the enforcing authorities may ignore these time limits.

Migrating Contamination

The provisions governing the escape of contaminating substances from one piece of land to another are contained within Section 78K of the 1990 Act.

This section makes it clear that where a person has caused or knowingly permitted substances to be in on or under land (as a result of which the land meets the definition of contaminated land) then that person (broadly the original polluter) is also responsible for the consequences should those substances migrate to adjoining sites. Therefore the original polluter is responsible not only for the remediation of his own land but also for the remediation of any other land which has been affected by such migrating contamination. The provisions of Section 78K go on to deal with the situation where the original polluter of the land from which the contaminants have escaped cannot be found.

In such a case the position is as follows. First, where contaminants have escaped from the source land to land A and the current owner/occupier of land A has not caused or knowingly permitted the substances to be on that land, then that current owner/occupier cannot be required to remediate any other land or water other than the land they occupy. In other words, the owner/occupier of land A will not be liable under these provisions to remediate any land other than his own (that is neither the source land nor other land to which the contaminants have escaped) unless either the owner/occupier of land A has caused or permitted the further escape or the owner/occupier of land A is also the owner/occupier of land B.

Secondly, in a situation where person A has caused or knowingly permitted land A to be contaminated and then sells that land to person B and contaminants then migrate from the land to elsewhere, the current owner/occupier of land A (namely person B) cannot be required to remediate the land to which the substances have escaped except to the extent that person B caused or knowingly permitted the escape. Person B is however responsible in the absence of the original polluter (the person they bought from) for remediating the site he now owns.

Knowledge in this situation is highly relevant because where a purchaser acquires a contaminated site and as a result of making enquiries finds out that the site to be acquired is also contaminating adjoining land, then by acquiring the site the purchaser can become an appropriate person for clean-up purposes (assuming the original polluter cannot be identified) due to knowing about the ongoing contamination of the adjoining land and taking no remedial action.

The Guidance provides specific advice on the phrase "caused or knowingly permitted", and this will be further considered in Chapter 6.

The provisions concerning migrating pollution are perhaps better understood if illustrated as follows:

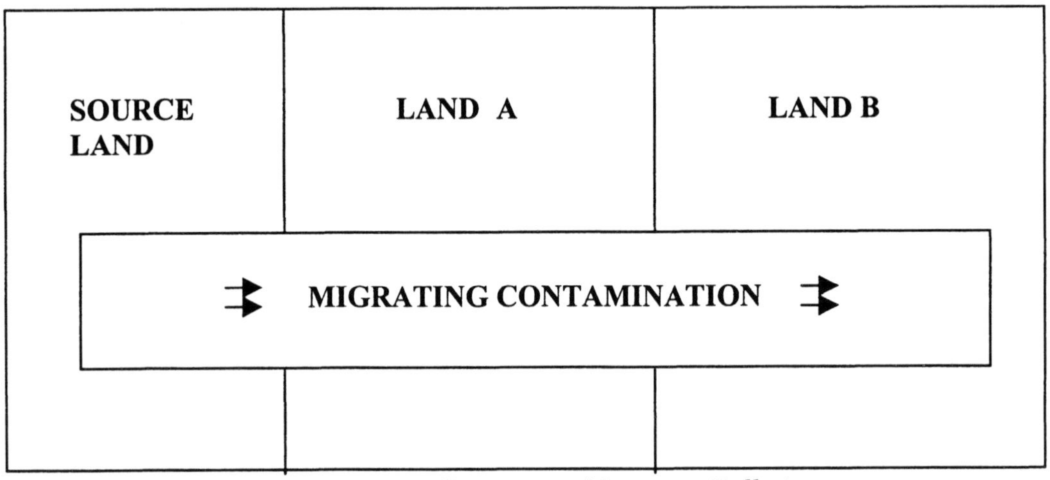

Figure 4.1 *Provisions Concerning Migrating Pollution*

The effect of Section 78K is as follows:

- If the person who caused or knowingly permitted the source land to be contaminated can be identified as an appropriate person then that person is responsible not only for remediating the source land but also land A and land B;
- In the event that the original polluter of the source land cannot be found then the innocent owner of land A is liable to remediate his own land but is not liable to remediate either the source land or land B unless he has either caused or knowingly permitted the contamination in question or is also the owner/occupier of the source land or land B as the case may be;
- If the owner of the source land sells it to a purchaser "P", P will not be liable to remediate land A or land B unless P caused or knowingly permitted the escape. If however, at the time that a remediation notice is served the original polluter who sold the land to P cannot be found then P will be liable in his capacity as owner/occupier for remediating the source land that he purchased.

As a result of these provisions it will be even more important than previously to conduct proper enquiries prior to the acquisition of a site which could meet the definition of contaminated land as it is quite clear from section 78K, that in the situation where the original polluter cannot be found current innocent owners/occupiers will be regarded as appropriate persons in defined circumstances even though they will not be regarded as fully responsible for the consequences of an ongoing migration of contamination as the original polluter would have been.

Appeals Against Remediation Notices

Section 78L of the 1990 Act contains the appeal mechanism in respect of remediation notices. It is possible to appeal against a remediation notice within 21 days of the notice being served and the 2000 Regulations specify 19 grounds upon which remediation notices can be appealed, which include:

(a) failure by the enforcing authority to have regard to the Guidance;
(b) insufficient time to carry out the works being provided by the notice;
(c) some informality, defect or error in the notice.

The appellate body (that is the person to whom the recipient appeals) is different depending upon which enforcing authority is taking the action. Where the local authority serves the notice then an appeal is to the Magistrates Court, where the Environment Agency serves the notice the appeal is to the Secretary of State for the Environment. On appeal the Magistrates Court or the Secretary of State can either quash the notice or confirm the notice with or without modification.

Careful consideration should be given to the opportunity to appeal, as not only will the recipient of the notice be responsible for the remediation, but as stated earlier, failure to comply with a remediation notice is a criminal offence with consequent financial penalties.

If however the enforcing authority believes that such criminal proceedings would be an "ineffectual remedy" against the person who has failed to comply with a remediation notice then a High Court Order can be obtained. Failure to comply with such an Order is a criminal contempt and could in extreme cases lead to an unlimited fine or imprisonment or both.

Non-compliance with a remediation notice has, however, more wide-ranging consequences than enforcement action. Section 78N of the 1990 Act contains the regime under which the enforcing authority can in appropriate cases undertake remediation work itself, which includes the situation where an appropriate person fails to comply with a remediation notice. (See Chapter 5).

The procedures in relation to the service of a remediation notice and enforcement action are perhaps best illustrated in Figures 4.2 and 4.3.

Summary: What Does It Mean For You?

Due to the widely-drawn definition of remediation it is necessary to move away from the view that it is simply a case of undertaking "pure clean-up work" and to accept that there may be a cost burden both before and after such work has been completed.

The provisions relating to migrating contamination make it essential that where a site is being purchased which has had a variety of industrial uses over the years the site should be thoroughly surveyed, if necessary with the benefit of environmental consultants, to determine not only the existing state of the land but also to assist in evaluating any prospective liabilities due to pollutants entering neighbouring land.

Further, where a site which is being acquired has not had contaminating uses itself, the state of adjoining sites should be investigated as far as possible to determine whether there is a risk that it might have become contaminated due to migrating contamination from those sites.

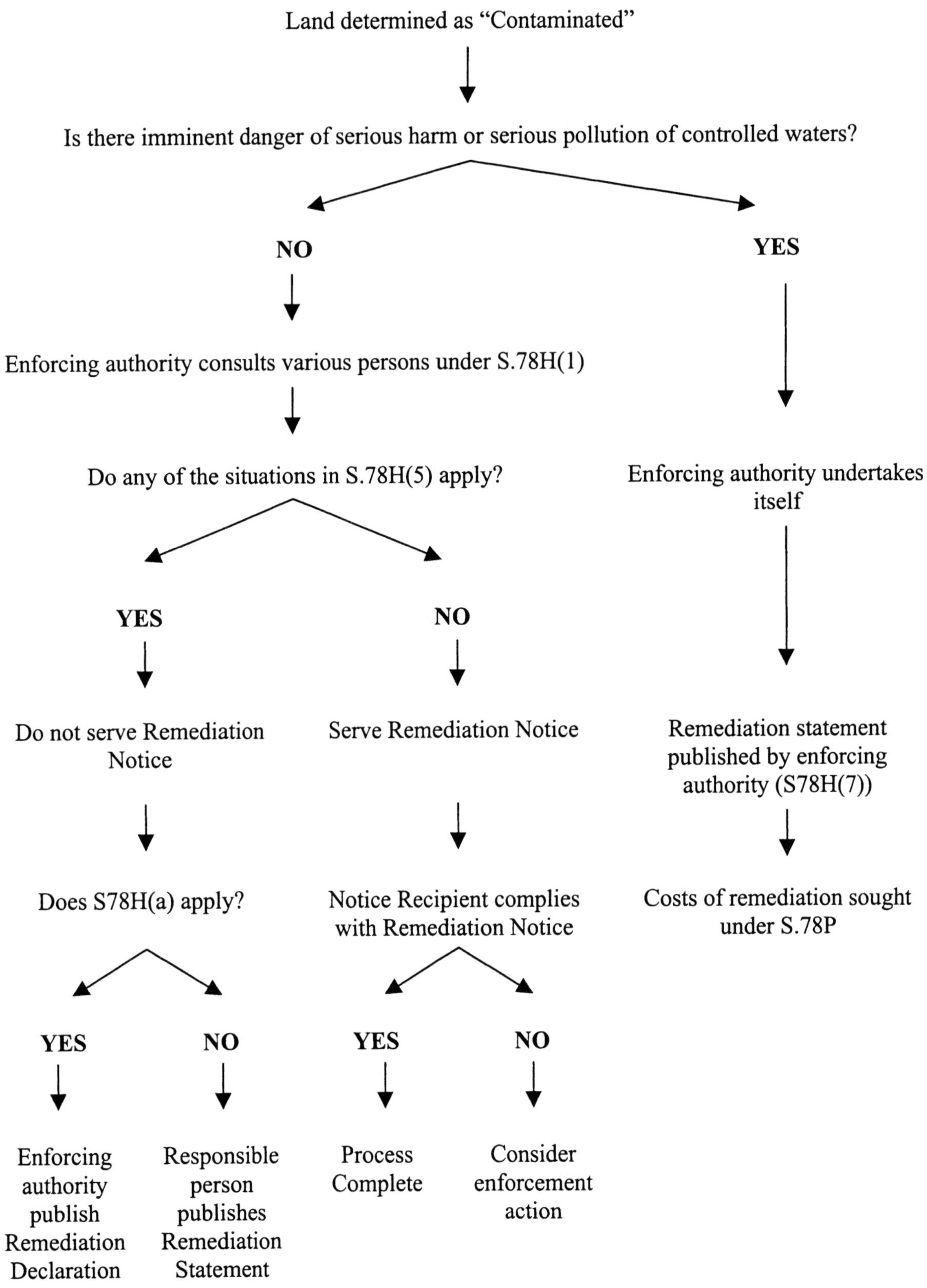

Figure 4.2 *Remediation Notice Procedure*

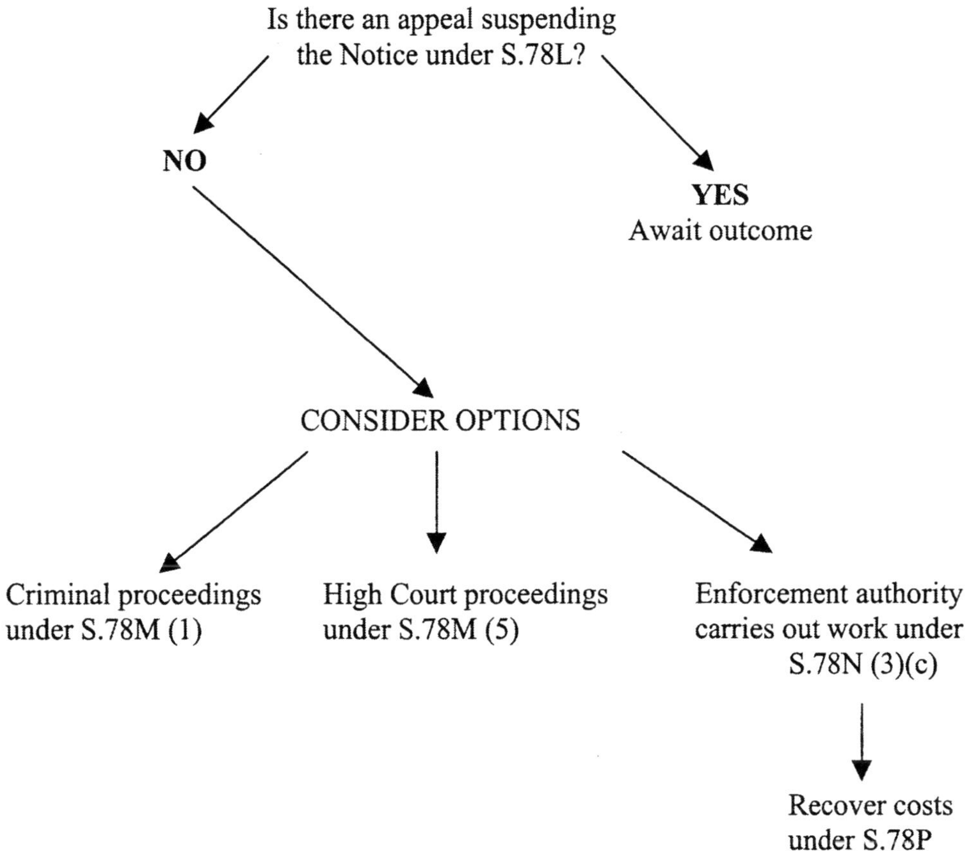

Figure 4.3 *Non-compliance with a Remediation Notice*

The cost benefit argument is a strong weapon in the armoury of any potential recipient of a remediation notice as, for example, if it can be suggested that remediation would cost in excess of £1 million and the existing contamination is not sufficiently serious to warrant that type of expenditure, which will call into question whether or not it is reasonable for the enforcing authority to proceed with enforcement action.

CHAPTER 5
THE TARGET FOR CLEAN-UP COSTS: APPROPRIATE PERSONS

Introduction

Assuming that the statutory definition of contaminated land has been met and an appropriate standard of remediation has been determined, the next and perhaps most crucial issue which arises is to identify the person upon whom a remediation notice or remediation notices should be served. At the very heart of the contaminated land provisions is Section 78F of the 1990 Act, which contains the statutory definition of the appropriate person or persons for the purposes of serving remediation notices. The definition is contained at subsections 2 and 4 that state as follows:

> *"(2) Subject to the following provisions of this Section, any person, or any of the persons, who caused or knowingly permitted the substances, or any of the substances, by reason of which the contaminated land in question is such land to be in, on or under that land is an appropriate person.*
>
> *(4) If no person has, after reasonable inquiry, been found who is by virtue of sub section (2) above an appropriate person to bear responsibility for the things which are to be done by way of remediation, the owner or occupier for the time being of the contaminated land in question is an appropriate person."*

This definition can be broken down as follows:

An appropriate person for the purpose of receiving a remediation notice from the appropriate enforcing authority is either:

- a person who caused the substances or any one of the substances to be in, on or under the land, (however such a person is only responsible for the consequences of the substances which they enabled to be in, on or under the land (section 78F (3)); or
- a person who knowingly permitted the substances to be in, on or under the land.

The above persons being referred to in the Guidance as Class A Persons, or:

- if after reasonable inquiry those who were effectively the original polluter or polluters (that is Class A Persons) cannot be found then the current owner or occupier.

The above person being referred to in the Guidance as a Class B Person.

There is a very important exemption in the definition of appropriate person in one specific situation, namely where a remediation notice requires clean-up work to be undertaken where land is contaminated land due to the second leg of the contaminated land definition being satisfied, that is, where the land is in such a condition as to cause either actual or potential pollution of controlled waters. In this case the definition of appropriate person is narrowed. Whilst those who caused or knowingly permitted the contaminants to be in, on or under the land so as to pose the actual or potential threat of pollution to controlled waters, can be made liable by being served with a remediation notice, the current owner or occupier of such land cannot be (Section 78J).

However the current owner or occupier can face potential liability for clean-up under Section 161 of the Water Resources Act 1991, which enables a "works notice" to be served in suitable circumstances. This will be further considered in Chapter 8.

Where it is established that an appropriate person caused or knowingly permitted a specific chemical to be on site which led to the site's contamination, that person is deemed to be the appropriate person even if that chemical has now changed as a result of some further chemical/biological process affecting it (Section 78F (9)).

As an example of how the appropriate person regime works, imagine a site which has been purchased from "X" by "Y" and has subsequently been determined to contain three contaminants, only two of which can be linked to "X"'s use of the site. "X" will be the appropriate person for the two as a Class A Person and "Y" will be the appropriate person for the other as a Class B Person (subject to provisions which are considered in Chapter 8).

Class A Persons and Causing or Knowingly Permitting Contamination

The primary category of appropriate person is the person or persons who caused or knowingly permitted the contamination to occur. The phrase "cause or knowingly permits" is a well-known phrase within water law and is currently found at Section 85 of the Water Resources Act 1991. In that Section it forms the main prosecution power of the Environment Agency when they take action in respect of water pollution, specifically causing or knowingly permitting poisonous, noxious or polluting matter to enter controlled waters.

As a result of Environment Agency prosecution activity in recent years and actions that predate its inception, there is now a substantial body of case law as to the meaning of the word "cause".

It was established by the House of Lords decision in *Alphacell Limited* v *Woodward (1972)* that as long as there is a link between the acts of the accused and the occurrence of the poisonous, noxious or polluting matter entering the controlled water then the prosecution is likely to succeed. In that case a paper pulp plant had a series of settling tanks connected by pipes, through which pulp-laden water was pumped. The pulp contained in the water settled out in the tanks after which the cleaned water entered a nearby river. When the system broke down due to the blockage of the pump inlets by vegetation, untreated water entered the river and the Company was successfully prosecuted for "causing" pollution. The House of Lords said:

> *"What then is meant by the word "caused" in the subsection? If a man intending to secure a particular result, does an act which brings that about, he causes that result. If he deliberately and intentionally does certain acts of which the natural consequence is that certain results ensue, may he not also be said to have caused those results even though they may not be intended by him? I think he can, just as he can be said to cause the result if he is negligent, without intending that result."*

In 1998 the case law was reviewed by the House of Lords *in Empress Car Co (Abertillery) Ltd* v *National Rivers Authority*, where their Lordships concluded that there was no requirement on the prosecution to prove the defendant did something which was the immediate cause of the pollution; the maintaining of tanks, lagoons or sewage systems full of noxious liquid was doing something, even if the immediate cause of the pollution was lack of maintenance, a natural event or the act of a third party.

It follows from both historic and recent case law that in the context of causing land to be contaminated, all a local authority has to be able to prove is that the actions of the appropriate person led to the presence of the substance, even if the appropriate person was not negligent and had no intention of causing the contamination in question. The Guidance reflects this approach:

> *"In the Government's view, the test of 'causing' will require that the person concerned was involved in some active operation...to which the presence of the pollutant is attributable. Such involvement may also take the form of a failure to act in certain circumstances."*

By contrast the phrase "knowingly permit" has not generated a similar volume of case law in the environmental context, principally because of the fact that to secure a conviction of knowingly permitting under the water regime it was always the case that knowledge had to be shown on the part of the accused and for this reason the charge was rarely relied upon. One of the few reported cases however on "knowingly permit" is the case *of Impress (Worcester) Ltd* v *Rees (1971)*, where the valves on oil storage tanks on the company's site were opened leading to oil polluting a nearby river. The company were prosecuted for knowingly permitting the pollution but the case was dismissed on the basis that as the valves were opened at night by a trespasser the event was outside the knowledge of the company and they could not therefore be said to have knowingly permitted the pollution.

More recently, the House of Lords considered the definition of "permitted" in the context of the tachograph legislation in *Nutall* v *Vehicle Inspectorate (1999)*, and stated that it was capable of bearing a wide interpretation including the act of not taking reasonable steps to prevent something which it was within one's power to prevent. In the case in issue they included the failure by the employer to examine the drivers' records to ensure they were complying with the legislation.

Thus, in a situation where it is alleged that an appropriate person is an appropriate person because they "knowingly permitted" the pollution, such an individual will wish to challenge the degree of knowledge being attributed to him and will also wish to argue that he took all possible steps to prevent the contamination from being in/on/under the land. It is likely therefore that enforcing authorities will rely mainly upon the first leg of the statutory definition, namely "causing".

Class B Persons

If it is not possible to identify the person who caused or knowingly permitted the contamination or if they can be identified they no longer exist, the current owner or occupier takes the stage. "Owner" is defined in the 1990 Act in relation to England and Wales as a person other than a mortgagee not in possession (that is, other than a lender who has not repossessed the site as a result of mortgage arrears) who whether in his own right or as a trustee for another person is entitled to receive the rack rent of the land or, where the land is not let at a rack rent, would be so entitled if it were so let. What this means is that anyone or any company owning land on a freehold basis (whether or not subject to a mortgage) or on a long leasehold basis will be an owner for the purposes of the legislation.

This definition of owner at first sight is problematic for the major lending institutions that lend substantial sums to industry as, in the event of default, should they take possession of the site it is almost certain that they will become an "owner". Thus in future these institutions will need to carefully consider whether it is worth taking possession of a contaminated site or whether it is safer to "walk away" from it and write-off the arrears.

However in the event that the Official Receiver or insolvency practitioner is appointed, Section 78X (3) of the 1990 Act becomes relevant. This provision effectively provides an immunity for the Receiver/insolvency practitioner in respect of any contamination in, on or under the land with which they are dealing and allows them to ignore the provisions of any remediation notice, unless the contamination is due to an act or omission by them which is unreasonable.

Whilst a significant defence is provided by this section it is perhaps not a complete defence, particularly where contamination is continuing on a site, as in such a situation it might be argued that it is indeed unreasonable for the Receiver/insolvency practitioner to take no action and thus challenge their immunity.

By contrast the word "occupier" is not defined by the legislation and it will therefore be given a factual commonsense meaning so that "occupier" means the person or company in physical occupation of the land in question.

Determination Process

When determining who the appropriate persons are and their respective liabilities the enforcing authority must act in accordance with the Guidance. The Guidance provides a five-stage process for determining such liabilities namely, in chronological order:

- Identifying Potential Appropriate Persons and Liability Groups;
- Characterising Remediation Actions;
- Attributing Responsibility between Liability Groups;
- Excluding Members of Liability Group;
- Apportioning Liability between Members of a Liability Group.

The Guidance provides detailed advice on each stage of the process and this will be considered in the next chapter.

Entry onto Third Party Land

There may be occasions when an appropriate person needs to enter onto land not in his ownership in order to remediate. In such circumstances the 1990 Act provides that the third party must consent to the appropriate person doing the work but is entitled to be duly compensated. Section 78G provides the power of entry with the compensation provisions being contained within Schedule 2 to the 2000 Regulations. Compensation is available for a variety of items including, depreciation and disturbance, and is payable to the mortgagee (if there is one).

Clean-Up by Enforcing Authorities

Section 78N of the 1990 Act gives the enforcing authority a power to undertake remediation work itself in a number of situations:

(a) Where the authority considers it necessary to prevent the occurrence of any serious harm or serious pollution of controlled waters of which there is an imminent danger;
(b) Where an appropriate person has entered into a written agreement with the enforcing authority for the authority at the cost of the appropriate person to undertake the work;
(c) (As mentioned at Chapter 4 above) where an appropriate person fails to comply with a remediation notice duly served;
(d) In circumstances first where the enforcing authority cannot serve a remediation notice in respect of pollution of controlled waters due to the provisions of section 78J and secondly, where the enforcing authority cannot serve a remediation notice due to the restrictions in relation to migrating contaminants contained at Section 78K;
(e) Where the enforcing authority considers that to recover either all or a proportion of the clean up costs would lead to "hardship" to the person(s) from whom it would be entitled to recover as defined by section 78P (see below);
(f) Where the enforcing authority cannot, despite reasonable inquiry find an appropriate person in relation to a specific thing, that is, in relation to a specific contaminant where say, there are three chemicals on the site left by three appropriate persons but only two have been found.

Cost Recovery

The 1990 Act envisages that where the enforcing authority has undertaken remediation work it should (subject to the inevitable exceptions!) be able to recover its costs from the appropriate persons responsible. The cost recovery regime is contained at Section 78P of the Act which makes it clear that in deciding whether to recover costs, and if so what proportion of the costs, the enforcing authorities must have regard to any hardship which the recovery may cause to the person who would have to pay the bill and also to the Guidance issued by the Secretary of State.

The Guidance provides that in making any cost recovery decision the enforcing authority should have regard to the following principles:

(a) The authority should aim for an overall result which is as fair and equitable as possible to all who may have to meet the costs of remediation, including national and local tax payers;
(b) The "polluter pays" principle, by virtue of which the costs of remediating pollution are to be borne by the polluter; the authority should therefore consider the degree and nature of responsibility of the appropriate person for the creation, or continued existence, of the circumstances which led to the land in question being identified as contaminated land.

Where the enforcing authority is looking to recover costs from a person who is the owner of any premises which include or consist of contaminated land and who caused or knowingly permitted the contamination to be on the land, then the authority can (if payment is not immediately forthcoming) serve a "charging notice" on that person which specifies the amount which the enforcing authority is looking to recover and the rate of interest from the date of service of the notice to the date of settlement. Indeed the Guidance suggests that the enforcing authorities should consider deferring recovery of their costs by way of such a notice if they believe they would recover more of their costs by so doing.

There is a right of appeal against the service of such a notice to the county court and, in the absence of a successful appeal, if the monies due are not paid the enforcing authority can in extreme cases require the site to be sold to discharge the debt. In other words non-compliance with a charging notice is effectively the same as allowing mortgage arrears to accumulate where the lender has very similar powers.

The 2000 Regulations do not, unfortunately, provide the fine detail of "charging notices" nor the appeal rights, thus leaving the county court to determine what grounds of appeal it will accept.

Cost Recovery Policy

The Guidance suggests that local authorities should prepare a publicly available document explaining their decision-making policy in respect of cost recovery;

> *"In order to promote transparency, fairness and consistency, an enforcing authority which is a local authority may wish to prepare, adopt and make available as appropriate a policy statement about the general approach it intends to follow in making cost recovery decisions. This would outline circumstances in which it would waive or reduce cost recovery (and thereby, by inference, not serve a remediation notice because it has the powers to carry out the remediation itself), having had regard to hardship and the statutory guidance."*

This would clearly be a document that should be actively sought out and requested by any likely recipient of a remediation notice.

Registers

The whole process surrounding the contaminated land regime is intended to be public. Section 78R of the 1990 Act requires every enforcing authority to maintain a register containing details of the enforcement process. The register must contain details of the documents prescribed in the 2000 Regulations (Regulation 16 and Schedule 3) which includes the following matters:

- Details of remediation notices;
- Details of appeals against remediation notices;
- Remediation statements;
- Remediation declarations;
- Designation of "special sites".

The duty to maintain public registers is subject to the provision of Section 78S which allows information relating to national security to be excluded from the register and Section 78T which allows commercially confidential information to be excluded from the register. In relation to confidential information the Act allows an appeal to the Secretary of State if an enforcing authority refuses to agree that the information in question is indeed commercially confidential.

The registers, which are to be kept at the local authority's principal office or the Environment Agency's regional office as appropriate, are to be available at all reasonable hours for inspection by the public, free of charge and copying facilities must be made available at a reasonable charge (Section 76R and Regulation 16).

Reports

The Environment Agency is given a co-ordinating role in preparing and publishing reports on the state of contaminated land and it is likely that it will publish such reports on at least an annual basis. This should provide an interesting overview of local authority activity across England and Wales as well as Agency activity in respect of special sites (Section 78U).

It will be appreciated that the regime outlined in this chapter provides a new and comprehensive structure within which local authorities and the Environment Agency will be able to tackle the problems arising from contaminated land on a site-specific, risk assessment based approach. Whether or not the regime is effective will depend not only upon the willingness of local authorities and the Environment Agency to undertake the task but also the inevitable issue of funding.

Summary: What Does It Mean For You?

The definition of appropriate person will become one of the most hotly disputed aspects of the new legislation given the financial obligations that follow from being classified as such. The effect of the definition is broadly to single out the polluter except where the polluter cannot be found when liability will fall to the owner/occupier.

In respect of commercial and industrial activities a view will need to be taken on whether the company is an original polluter having caused or knowingly permitted contamination or whether, on the basis of an acquisition by the company, it could be held to be an appropriate person due simply to possession. The benefit to the company of the latter is that it would only be potentially liable for remediation of it's own site and not any adjacent sites which have been affected by migrating pollution and then only if there was no Class A person available.

The concept of "knowingly permitted" is also potentially problematic because its ambit is uncertain. If it only applies to where original contamination is knowingly permitted, then the enforcement authorities will only be able to establish liability if they can prove knowledge on behalf of the party at the time of the contamination. There will be no liability based upon simple acquisition.

In the event however that a wider definition is given to this phrase and it is possible to knowingly permit simply by acquiring a site with knowledge, then it will become even more important than it currently is to ensure that environmental consultants reports are sought and legal advice taken.

The concept of appropriate person is described in Figure 5.1.

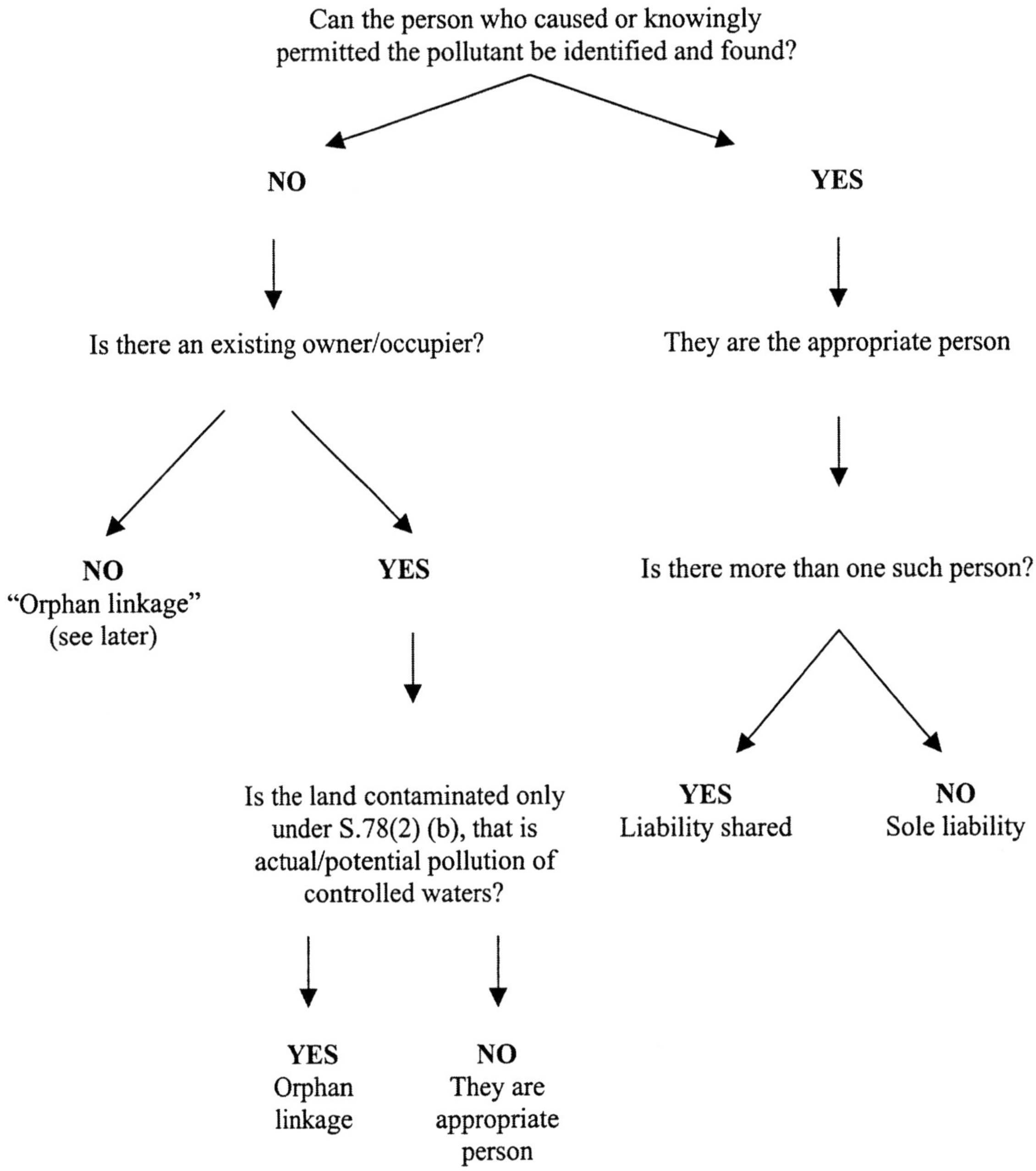

Figure 5.1 *Who is Responsible for Clean-up Costs?*

CHAPTER 6
CONTAMINATED LAND: CIVIL LIABILITY

Civil liability is the term used to describe the situation where one person or company sues another person or company in the civil courts for damages alleging that they have suffered some form of loss as a result, for example, of their land being contaminated by that other person. The terminology in the civil courts is that the Claimant is the person who brings the case alleging that they have been harmed by the activities of the Defendant. There are no statutes setting out a civil liability code in England and Wales, instead civil liability has evolved historically on the basis of decided cases.

In relation to contaminated land there were, prior to the House of Lords decision in *Cambridge Water Company* v *Eastern Counties Leather plc* (judgment pronounced on 9 December 1993), three main areas of potentially relevant civil liability in contaminated land cases. These were:

(i) liability in negligence;
(ii) liability in nuisance;
(iii) liability under the rule in *Rylands* v *Fletcher.*

Before looking at the current state of the law following the decision of the House of Lords, it is worth setting out briefly the legal basis of each type of claim.

(i) Negligence

There are a number of elements that must be satisfied to succeed in a claim in negligence:

- the Defendant owed the Claimant a duty of care;
- the Defendant breached the duty of care;
- the Claimant suffered damage of a kind that was foreseeable in type even if not in extent as a result of the breach.

The modern law of negligence can be traced back to the House of Lords decision in *Donoghue* v *Stevenson* in 1932. In that case the Claimant (or Plaintiff as they were then known) successfully sued a manufacturer who had supplied to a wholesaler who in turn had supplied to a retailer, a ginger beer drink which was contaminated due to the presence of a dead snail in the bottle! The House of Lords found that the manufacturer should have reasonably foreseen that a person consuming the drink might suffer harm and on that basis held that the manufacturer owed the final consumer a legal duty of care (even though there was no contractual relationship between the manufacturer and the final consumer). Having decided that a duty of care existed, the House of Lords decided that the duty had been fairly self-evidently breached and as a result the consumer had suffered foreseeable harm due to drinking the contaminated ginger beer.

Since that case there have been a series of cases extending the circumstances in which a duty of care is said to exist. By way of example, recent significant cases include *Spring* v *Guardian Royal Insurance* (1994), where the House of Lords held that in relation to the provision of a reference, an employer owed a duty of care not just to the former employee but also to the former employee's prospective employer in providing a proper and accurate reference, and also in 1994, the House of Lords held in *White* v *Jones*, that a duty of care was owed by a firm of solicitors to two intended beneficiaries with a result that the firm was liable when they delayed in executing a will on behalf of a client with the result that the client died before the will was finally formalised.

In relation to contaminated land liability, to pursue a claim successfully in negligence, the Claimant would have to show firstly that the Defendant owed a duty of care to him. In practice this should not be too difficult to establish as it is reasonably foreseeable that if, for example, chemicals were allowed to escape from the Defendant's site to the Claimant's site, the Claimant would suffer physical damage to his property as a result of the contamination. Thus, a duty will be imposed.

It is likely therefore, that having established a duty of care a Claimant will also be able to show a breach of the duty and will thus be able to recover damages, so long as the loss suffered was foreseeable in principle even if it was not foreseeable in terms of the full extent of the actual loss suffered. In relation to compensation there continue to be problems in relation to pure financial (or economic) loss that does not arise directly from some damage as opposed to compensation for the physical damage itself. The law has always allowed recovery for physical damage and for personal injury, but traditionally has taken a more restrictive approach to economic loss.

On the basis of current case law it is therefore likely that where land is contaminated, a Claimant, if able to successfully show breach of a duty of care, will be able to recover damages for the physical cost of cleaning up the land and possibly for any loss of profit (for example, loss of manufacturer's time) suffered as a result of the contamination.

(ii) Nuisance

The term nuisance covers a wide variety of acts or omissions and deals with damage or interference to property or interests in property. There is an overlap between nuisance and negligence but they are essentially different: in negligence it is not necessary to own land to recover damages whereas in nuisance, what is known as an interest in land must be held by the Claimant, that is, the Claimant must have a freehold or leasehold interest in the land to be able to recover damages.

At common law, there are two types of nuisance - public and private. Public nuisance is a situation involving damage or obstruction to a group of the public, for example an obstruction of the highway that affects all those needing to use the highway. It is not possible however to recover damages in public nuisance unless the Claimant can show that he has suffered special damage which is damage over and above that suffered by the public at large.

By contrast, under the law of private nuisance, a Claimant can recover damages where his private property rights have been interfered with without having to show any special damage. In private nuisance following the House of Lords' decision in *Cambridge Water* there are two broad hurdles for a Claimant to jump:

(a) First the Claimant must show that as a matter of fact (the legal term being "as a matter of causation") the activities of the Defendant actually caused the damage to the Claimant's property. In relation to contaminated land, it must be proved on the balance of probabilities that the chemicals present on the Claimant's land as a matter of fact seeped onto the land from the Defendant's property and the contamination caused loss or damage;

(b) Secondly, having proved as a matter of fact that damage has occurred, a Claimant must be able to show that it was foreseeable to the Defendant at the time he (the Defendant) was conducting the activities which led to the contamination, that the consequence of those activities could be the resulting contamination. This was the major issue before the House of Lords in the *Cambridge Water* case.

In nuisance, if the Claimant jumps both the causation and foreseeability hurdles then he will recover damages for all foreseeable losses which have been occasioned as a result of the Defendant's activities.

Although not part of the law of nuisance, it is perhaps convenient to mention at this point what are known as riparian rights. Under the common law the owner of a piece of land on the banks of a natural stream or river has particular rights known as riparian rights. The Courts, in two old cases, *Chasemore* v *Richards (1659)* and *Young & Co* v *Bankier Distillery Company (1893),* made clear that those with adjoining watercourses have a right to enjoy the water in a pure and natural state and must do nothing to the water which adversely affects riparian owners downstream who have exactly the same rights. This area of law has not been used much in recent years, but in a situation where land adjoining a rivers pollutes it, a downstream owner could pursue this argument along with more traditional arguments based on negligence and nuisance.

(iii) Liability under the Rule in *Rylands* v *Fletcher*

The third area of civil liability, and regarded as a separate area in its own right prior to the House of Lords' decision in *Cambridge Water*, was what had become known as the Rule in *Rylands* v *Fletcher* after a case of that name decided in 1868.

In that case the Defendant had built a reservoir on its land and filled it with water. The water later escaped down old mineshafts under the land which had not been effectively blocked-up and flowed through those shafts onto the Claimant's land flooding the Claimant's mine.

The case historically had been believed to have created a separate head of liability whereby if something was brought onto land that wasn't naturally there and there was an escape of that thing causing damage, liability was said to be strict. In other words the Defendant could be liable to the Claimant even if the Defendant did not foresee the damage occurring to the Claimant as a result of his acts or omissions. The House of Lords in *Rylands* v *Fletcher* said:

> *"If a person brings or accumulates on his land anything which, if it should escape, may cause damage to his neighbour, he does so at his peril. If it does escape and cause damage he is responsible, however careful he may have been, and whatever precautions he may have taken to prevent the damage."*

The rule over the years was narrowed with a distinction being drawn by the Courts between a "natural use" of the land on the part of the Defendant which led to the Defendant not being liable and a "non-natural use" on the part of the Defendant which led to his being liable.

A Claimant who had suffered land contamination, therefore, would naturally be tempted to pursue a claim under *Rylands* v *Fletcher* because if an escape could be argued to have taken place and if the courts were persuaded that the use of the land on the part of the Defendant was non-natural then the foreseeability issue would be neatly circumvented. The appropriateness of this head of claim was considered by the House of Lords in the *Cambridge Water* case.

Cambridge Water Company* v *Eastern Counties Leather plc

The House of Lords in *Cambridge Water Company* v *Eastern Counties Leather plc* reviewed the law of nuisance and the rule in *Rylands* v *Fletcher* in the context of a case of historic pollution. The law of negligence was not reviewed by the House of Lords as arguments based on negligence were not pursued after the High Court hearing.

It is worth setting out the facts in some detail given the significance of the House of Lords' decision. Cambridge Water Company ("CWC") operated a borehole at the village of Sawston, near Cambridge. About 1.3 miles away from the borehole Eastern Counties Leather plc ("ECL") operated the business of a tannery producing a range of fine leather goods and had occupied the site since 1879. To enable it to degrease pelts ECL began using a particular type of solvent (PCE) in the 1960's and continued to use that solvent until 1991. From the early 1960's until 1976 PCE was delivered in 40-gallon drums and when it was needed a drum was taken by forklift truck to the relevant machine and tipped into a tank at the base of the machine with the result that there was considerable spillage. From 1976 until ECL stopped using PCE in 1991, the solvent was

CHAPTER 8
CONTAMINATED LAND AND WATER PROTECTION

As stated in Chapter 3, the definition of "contaminated land" in Section 78A (2) of the 1990 Act, clearly acknowledges the specific potential impact of contaminated land upon controlled waters. However, during the passage of the Environment Bill through Parliament, the National Rivers Authority (NRA) (the Environment Agency's predecessor) had expressed serious concerns about the deficiencies in, and practicability of, what was being proposed and some of those concerns remain. Despite the Guidance, there is considerable ambiguity as to what constitutes "significant harm" in the first limb of the Section 78A (2) definition and what criteria will satisfy the definition of "serious harm" which determines whether or not a site is a "special site" and which therefore falls within the control of the Agency rather than the local authority. Furthermore, there are problems associated with, in effect, the inability of the local authority/Agency to serve remediation notices in respect of groundwater (or indeed surface waters) which have been contaminated by an unknown party. In addition to these issues, the Agency will have to resolve how it interprets the interface between the contaminated land provisions in the 1990 Act and the enhanced clean-up powers which it has been given under Section 161A of the Water Resources Act.

Site Designation

Whilst the current definition of contaminated land is clear in relation to the direct effect upon the water environment (that is, where "pollution of controlled waters is being, or is likely to be, caused") there is, as previously stated, considerable scope for ambiguity over the definition of "significant harm" with respect to the soil of the land which could ultimately impact on groundwater. The Guidance assists to some degree but the definitive position will no doubt have to evolve, together with Environment Agency and local authority policy, with the assistance of the Courts!

In respect of water pollution the 2000 Regulations provide for the designation of land as a "special site" where:

- controlled waters which are intended to be used for the supply of drinking water are being affected by the land;
- controlled waters do not or are not likely to meet the criterion for relevant classification under the Water Resources Act 1991 due to the land;
- any of the substances contained in Schedule 1 to the Regulations are the source of the pollution (this includes cyanide, mercury and mineral oil) of the waters;
- any part of the waters are contained within underground rock strata of the types specified in Schedule 1 of the Regulations.

Where any of these situations exist the Agency will take responsibility for the site but the initial determination will rest with the local authorities (subject to the right of the Agency to give notice to a local authority that they require a specific site to be designated under Section78C (4)). It is they who will inspect their area "from time to time" for the purpose of identifying contaminated land, and those sites falling within the categories specified in the 2000 Regulations will be designated as special sites and become the responsibility of the Agency. Local authorities must seek the advice of the Agency and have regard to such advice when making the determination. There is a mechanism for resolving disputes between the Agency and local authorities as to whether a site is "special", whether that determination comes from the Local Authority (Section 78C (1)) or the Agency (Section 78C (4)), with the referral of the decision to the Secretary of State for final determination. This appeal mechanism should be seen as a "last resort" as it would

> *in negligence. For if a Plaintiff is in ordinary circumstances only able to claim damages in respect of personal injuries where he can prove such foreseeability on the part of the Defendant, it is difficult to see why, in common justice, he should be in a stronger position to claim damages for interference with the enjoyment of his land where the Defendant was unable to foresee such damage."*

There has for many years been a debate as to the requisite standard for liability with the question being posed as to whether liability should be fault or foreseeability based, that is whether liability should be effectively strict without the need for the Claimant to prove fault on the part of the Defendant, and the House of Lords addressed this issue in the speech of Lord Goff:

> *"...I incline to the opinion that, as a general rule, it is more appropriate for strict liability in respect of operations of higher risk to be imposed by Parliament than by the Courts. If such liability is imposed by statute, the relevant activities can be identified and those concerned can know where they stand. Furthermore, statute can where appropriate lay down precise criteria establishing the incidence and scope of such liability.*
> *It is of particular relevance that the present case is concerned with environmental pollution. The protection and preservation of the environment is now perceived as being of crucial importance to the future of mankind; and public bodies, both national and international, are taking significant steps towards the establishment of legislation which will promote the protection of the environment, and make the polluter pay for damage to the environment for which he is responsible...But it does not follow from these developments that a common law principle, such as the rule in Rylands v Fletcher, should be developed or rendered more strict to provide for liability in respect of such pollution. On the contrary, given that so much well-informed and carefully structured legislation is now being put in place for this purpose, there is less need for the courts to develop a common law principle to achieve the same end, and indeed it may well be undesirable that they should do so."*

The House of Lords in giving its judgment in *Cambridge Water*, was well aware of the fact that the Government was in the middle of its wide ranging review of civil liability for contaminated land, with the Environment Act 1995 subsequently representing the "carefully structured legislation" it anticipated.

The principles of the House of Lords decision in *Cambridge Water* have been applied in the High Court in the case of *Graham* v *Rechem International Ltd* (judgment given 16th June 1995). In this case, Graham was a farmer who claimed that his pedigree Ayrshire cows had been poisoned as a result of Rechem's toxic waste incineration plant at Bonnybridge in Scotland, leading to the collapse of his dairy business. He claimed damages amounting to £1.5 million under the heads of negligence and nuisance. He failed in his claim because as a matter of causation he did not establish that the emissions from the Rechem plant contributed or materially contributed to the damage alleged to have been suffered. However, if the causation hurdle had been successfully jumped it is likely that Graham would have received damages as the Court held that liability in such a case was effectively strict with the Defendant being unable to avoid liability by arguing that they had taken all reasonable care as the incineration of the chemical waste was held not to be a reasonable use of land.

The Court also confirmed that in relation to whether the Defendant could reasonably foresee the harm of the type that eventually occurred, it was not necessary to foresee the precise mechanisms by which the harm would result. In other words so long as damage to cattle was reasonably foreseeable Rechem would have been liable even though it could not foresee precisely how the animals would be harmed.

Thus, as a result of the *Cambridge Water* case, discovering the originating date of the pollution will obviously be of vital importance because what the Courts will regard as foreseeable in the 1960's and 1970's will obviously be very different from what the Courts will regard as foreseeable in the 1990's.

Having considered the civil position at some length, it is evident that the practical advice must be to avoid having to take or defend such proceedings if at all possible as they are inevitably costly; the total bill in the *Cambridge Water* case exceeded £l million and the financial risk is multiplied by the fact that the loser pays the majority of the winners costs. Litigation should be seen as a last resort when all other attempts to resolve the matter have failed.

When contaminated sites are being disposed of as a commercial transaction, the issue of liability should be expressly dealt with and costs apportioned appropriately. This important practical issue is considered at Chapter 9.

As outlined at Chapter 2, the Government considered as part of its wide ranging review of contaminated land announced in March 1993, whether there should be any changes made to the common law rules by Acts of Parliament. The Paper "Paying for our Past" (March 1994) expressed the provisional view that:

> *"While therefore a cautious approach should be adopted towards any statutory changes to common law, its amendment should not be ruled out."*

It continued:

> *"It seems, subject to further analysis of the judgment in Cambridge Water, that the current mix of strict and fault-based approaches strikes a reasonable balance."*

The Government finalised its position in the Framework for Contaminated Land document (November 1994) by stating:

> *"The Government considers that the judgment (in Cambridge Water) clarifies the balance under the Common Law between private interests. On the one hand, a polluter or the owner of land from which pollution escapes should be liable for damage he causes to third parties' interests and should thereby be encouraged to behave responsibly and prudently. On the other hand, it is unfair to hold the polluter responsible for every consequence of his actions or omissions, no matter how unforeseeable the possibility of damage was at the time.*
> *The Government does not consider that the balance of advantage between the parties should be changed. Such suggestions as defences based on "the State of the Art", the application of due diligence, actions taken in good faith or in compliance with regulatory requirements would shift the balance. It is better to let the courts decide whether any of these is of evidential weight in a particular case than for Parliament to add to or reduce the rights of potential plaintiffs or defendants."*

There is a continuing debate at EU level as to the appropriate standard in civil cases. In 1993 the European Commission published a Green Paper on Environmental Liability which considered the advantages of a strict liability regime for historic pollution and detailed where such rules already operated within the EU. The Government formally responded opposing the EU view but it received partial support from a House of Lords Select Committee in December 1993.

In 1997 the EU published a working document which was intended to structure the debate and lead to the production of a White Paper. The White Paper was adopted by the European Commission on 9 February 2000 and whilst promoting a strict liability regime for damage caused by dangerous activities, it suggests a fault-based regime for damaged caused by non-dangerous activities. Neither regime will have retroactive application.

Summary: What Does It Mean For You?

It is important to appreciate that the common law rules discussed in this chapter are totally separate and distinct from the statutory regime contained in the Environmental Protection Act 1990. Thus, it is possible for a company to take action in the High Court alleging that a nuisance has occurred resulting in a site being contaminated and to recover damages in respect of that contamination, even though the site would not meet the statutory definition of contaminated land and even though the enforcing authority could not take remediation action in respect of the site.

Given the way the common law rules have evolved over time, it is very likely that the law of nuisance will provide a remedy in a wider range of cases than those cases where the enforcing authorities are in a position to act and may thus become an effective tool in contaminated land remediation.

Whether as a potential Claimant pursuing a claim or a potential Defendant defending a claim, the importance of technical evidence as to the actual state of contamination is extremely important. Although liability in nuisance is effectively strict once contamination has been proved (subject to foreseeability) many potential nuisance cases can fall at the first hurdle where the technical evidence relating to contamination is insufficiently conclusive.

CHAPTER 7
APPORTIONMENT AND EXCLUSION OF LIABILITY

In respect of any contaminated land site there will be a person or persons who will potentially bear liability for any significant pollutant linkage thereon and hence the remediation liability. In the Guidance these persons are referred to as the "liability group".

This group will be comprised of either Class A or Class B persons and the enforcing authority will have to determine if liability can be fixed to the "appropriate person" within that group and how such liability should be apportioned if there is more than one such person.

This determination will become more complex where there is a history of potentially polluting activities linked to the site and which may require a number of separate remediation actions to be undertaken.

As mentioned in the previous chapter, the Guidance provides a five stage procedure for making the determination and this will now be considered.

First Stage

As previously mentioned, the enforcing authority investigation should be to make reasonable enquiries to find those persons who have caused or knowingly permitted the pollutant in question to be in, on or under the land thereby creating the significant pollutant linkage. Such persons being the "Class A Liability Group". Where such persons cannot be found then the enforcing authority will identify the current owners or occupiers of the site as the "Class B Liability Group."

It should be noted that where no Class A or Class B persons can be found in respect of a site *or* where no Class A person can be found and the contamination issue arises solely due to the pollution of controlled waters, then the enforcing authority must treat that "significant pollutant linkage" as an "orphan linkage". This is quite simply as it suggests, a linkage where no appropriate person can be fixed with liability. In a case where the orphan linkage is the only pollutant linkage on the site, the enforcing authority itself will bear the cost of any remediation that is carried out. Where there are other significant pollutant linkages on the site, the liability for the orphan linkage will fall to either the Class A or Class B Liability Group that has been determined for that pollutant.

As mentioned previously, the contaminated site may well consist of a number of significant pollutant linkages, thus the enforcing authority must determine the Liability Group in respect of each such linkage.

The enforcing authority must then consider whether any members of the Liability Group are exempted from liability under any of the provisions of Part II A of the 1990 Act, for example where an innocent owner/occupier is exempted in relation to migrating contamination under Section 78K.

If by doing so all of the members of the Liability Group are exempted, the enforcing authority should treat that particular significant pollutant linkage as an orphan linkage.

Second Stage

This stage does not apply where there is a single significant pollutant linkage, however where there are two or more significant pollutant linkages, the enforcing authority must establish whether each remediation action is:

(a) referable solely to the significant pollutant in a single significant pollutant linkage (a "single linkage action"); or

(b) referable to the significant pollutant in more than one significant pollutant linkage (a "shared action").

Where a "shared action" exists, this is further divided by the Guidance into "common actions" and "collective actions" and the relationship is best explained by way of the following diagram:

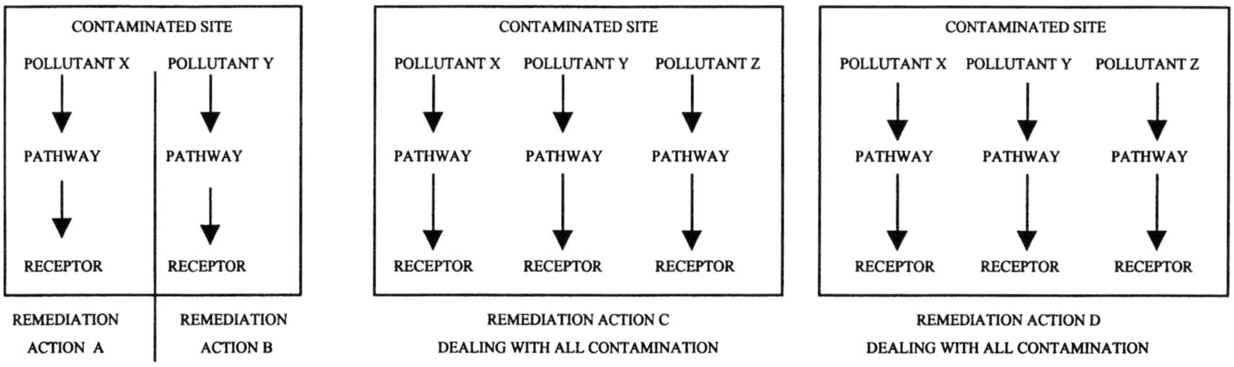

Figure 7.1 *Relationship Between Actions*

- the remediation actions A and B deal specifically and separately with the pollutants in two linkages and are therefore "single linkage actions" where the remediation is aimed at dealing with that single specific pollutant;
- if the remediation action proposed in C would still have been proposed if it were dealing with linkages X, Y and Z separately, it is a "common action";
- if the remediation action proposed in D would not have been proposed for X but would have been proposed for Y and Z (or vice versa) then it is a "collective action";
- the collective action therefore replaces individual remediation actions as it meets the standard of remediation required for all linkages.

Third Stage

Again this stage of the procedure does not apply to the situation where there is a single significant pollutant linkage and deals with attributing the responsibility between liability groups and refers to the case where there are "shared actions" as discussed previously. Where there is a "common action" then the following applies:

(a) if there is a single Class A Liability Group, then the full costs of carrying out the common action should be attributed to that Group, and no cost should be attributed to any Class B Liability Group;

(b) if there are two or more Class A Liability Groups then an equal share of the cost of carrying out the common action shall be attributed to each of those groups and no cost should be attributed to any Class B Liability Group;

(c) if there is no Class A Liability Group and there are two or more Class B Liability Groups, then the enforcing authority should treat those Liability Groups as if they formed a single Liability Group, attributing the cost of carrying out the common action to that combined group.

Where the situation creates a "collective action" then the same rules as stated above apply except that where the costs fall to be divided among several Class A Liability Groups, then instead of being divided equally the costs are apportioned by determining estimates of the hypothetical costs of each of the liability groups carrying out actions which are subsumed by the collective action.

Fourth Stage

Having identified the potential appropriate person and, where appropriate, characterised remediation actions and attributed responsibility between the Liability Groups, the enforcing authority must then consider whether any of the members of a Liability Group should be excluded.

The excluded categories are specified within the Guidance in relation to both Class A and Class B Liability Groups and is considered below.

Fifth Stage

By following the previous stages the enforcing authority will have a Liability Group upon whom to place responsibility for remediation. The next stage therefore is for the authority to determine how the costs of remediation should be apportioned between the members of that Group. Again the Guidance details how this apportionment process should occur in relation to both Class A and Class B Liability Groups and reference should be made to that Guidance.

Exclusion of Members of a Class A Liability Group

As stated above, the fourth stage of the liability determination process requires consideration of whether any members of the groups should be excluded.

There are six tests provided by the Guidance which are to be applied consecutively, but one of the overriding principles is that if as a result of applying a test no members of the Liability Group would remain, then that test should not be applied. In general terms the tests are as follows:

Test 1 - Excluded Activities

This test excludes those persons who are considered to have a link with the contaminated land which is not one which would justify a determination that liability should rest with them.

The list includes those providing financial assistance to another person (for example a mortgagee); those creating a tenancy over land in favour of a person who subsequently caused or knowingly permitted the presence of the significant pollutant linkage in question; and those issuing any statutory permission, licence or consent required for any action or omission by reason of which some other person appears to the enforcing authority to have caused or knowingly permitted the presence of the significant pollutants in question.

The list of excluded activities is included with the Guidance and reference should be made to that list for full details.

Test 2 – Payments made for Remediation

As the Guidance says:

> *"The purpose of this test is to exclude from liability those who have already, in effect, met their responsibilities by making certain kinds of payment to some other member of the Liability Group, which would have been sufficient to pay for adequate remediation."*

Such payments include payments which form part of a contract for the transfer of ownership of the land which is either specifically provided for in the contract to meet the cost of carrying out the remediation or which consists of a reduction in the contract price explicitly stated in the contract to be for that purpose.

In such circumstances the enforcing authority should exclude the person who made the payment in respect of the remediation action in question.

Test 3 – Sold with Information

This test is to exclude from liability those persons who despite having caused or knowingly permitted the presence of the significant pollutant in the land, have disposed of that land in circumstances where it is reasonable for another member of the Liability Group who acquired the land from them to bear liability for the remediation.

This applies to sales of land at arms length in full knowledge of the presence of the contamination in question.

Thus there is a clear incentive for the seller of land to make full disclosure to any purchaser as to the state of the land in question as this will enable the seller to rely on this exclusion test.

Test 4 – Changes to Substances

In a situation where persons become members of the Liability Group because they caused or knowingly permitted a substance in, on or under the land which by itself did not create a significant pollutant linkage but by interacting with another substance which was later introduced into the land did so, that person should be excluded from liability.

Test 5 – Escaped Substances

The purpose of this test is to exclude from liability those persons who would otherwise be liable for the remediation of contaminated land which had only become contaminated as a result of the escape of substances from other land, where it can be shown that another member of the Liability Group was actually responsible for that escape.

Test 6 – Introduction of Pathways or Receptors

This test excludes those persons who would be liable solely because of the subsequent introduction by another party of either a pathway creating the link between the pollutant and the receptor or the introduction of a receptor themselves.

For example if there is a contaminated site which does not fall within the definition of contaminated land within the Act because there is no receptor, but then somebody obtains planning permission for a housing development on that land thereby introducing receptors, and those receptors are then affected by the original contamination, then that person who initially caused or knowingly permitted the pollutant to be in the land should be excluded from liability.

It is important to note however that where such a development takes place on a neighbouring site, the test and exclusion does not apply.

Exclusion of Members of a Class B Liability Group

This test applies where two or more persons have been identified as Class B persons for a significant pollutant linkage. In such circumstances any such person who either occupies the land

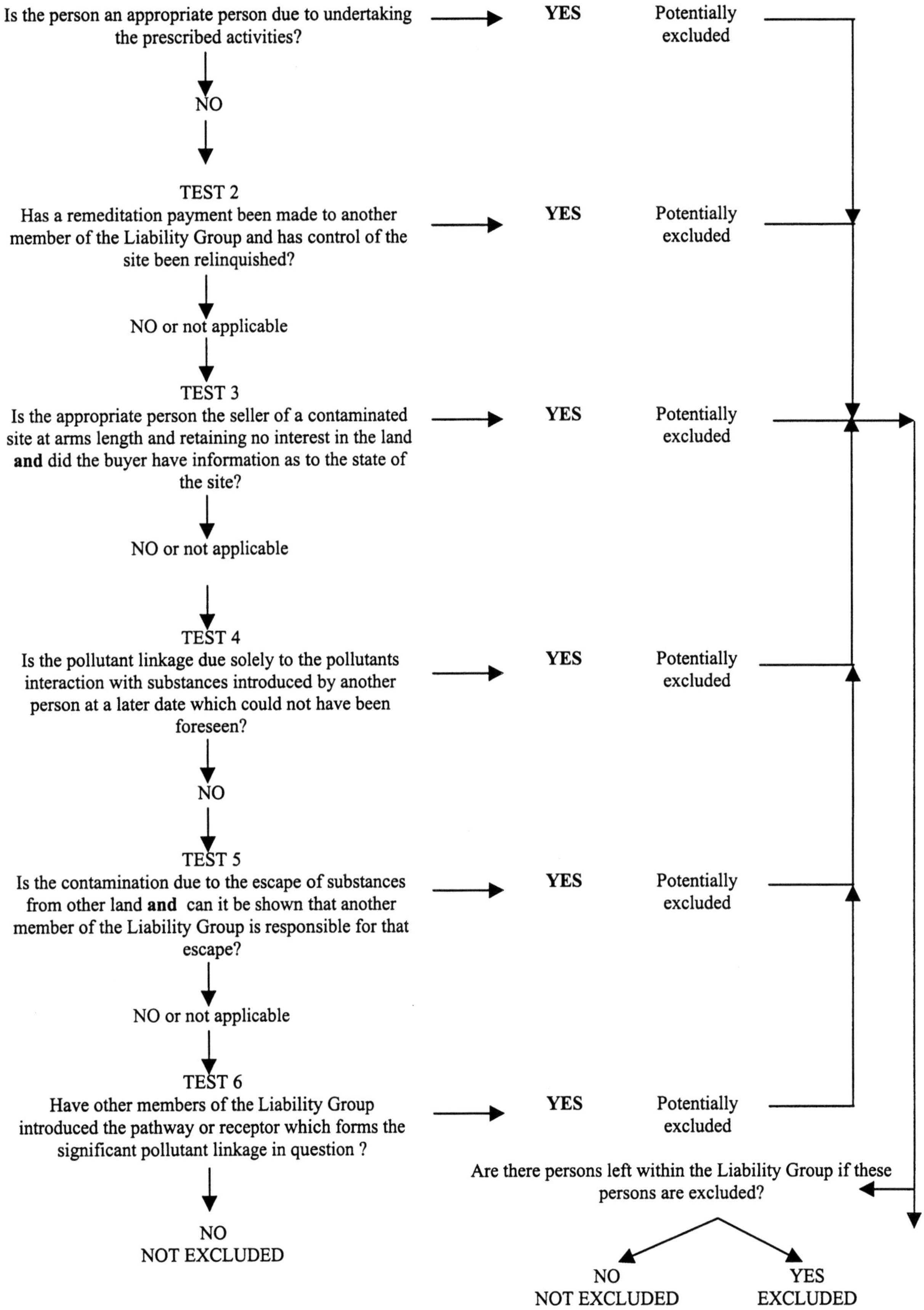

Figure 7.2 *Exclusion of Appropriate Persons: Class A*

under a licence or agreement which has no marketable value or which he is not legally able to assign or transfer to another person, or is liable to pay rent which is equivalent to the rack rent of such land and holds no beneficial interest in the land other than a tenancy to which the rent relates, should be excluded.

However, this test should not be applied and thus no exclusion made, if it would result in the exclusion of all members of the Liability Group.

CHAPTER 8
CONTAMINATED LAND AND WATER PROTECTION

As stated in Chapter 3, the definition of "contaminated land" in Section 78A (2) of the 1990 Act, clearly acknowledges the specific potential impact of contaminated land upon controlled waters. However, during the passage of the Environment Bill through Parliament, the National Rivers Authority (NRA) (the Environment Agency's predecessor) had expressed serious concerns about the deficiencies in, and practicability of, what was being proposed and some of those concerns remain. Despite the Guidance, there is considerable ambiguity as to what constitutes "significant harm" in the first limb of the Section 78A (2) definition and what criteria will satisfy the definition of "serious harm" which determines whether or not a site is a "special site" and which therefore falls within the control of the Agency rather than the local authority. Furthermore, there are problems associated with, in effect, the inability of the local authority/Agency to serve remediation notices in respect of groundwater (or indeed surface waters) which have been contaminated by an unknown party. In addition to these issues, the Agency will have to resolve how it interprets the interface between the contaminated land provisions in the 1990 Act and the enhanced clean-up powers which it has been given under Section 161A of the Water Resources Act.

Site Designation

Whilst the current definition of contaminated land is clear in relation to the direct effect upon the water environment (that is, where "pollution of controlled waters is being, or is likely to be, caused") there is, as previously stated, considerable scope for ambiguity over the definition of "significant harm" with respect to the soil of the land which could ultimately impact on groundwater. The Guidance assists to some degree but the definitive position will no doubt have to evolve, together with Environment Agency and local authority policy, with the assistance of the Courts!

In respect of water pollution the 2000 Regulations provide for the designation of land as a "special site" where:

- controlled waters which are intended to be used for the supply of drinking water are being affected by the land;
- controlled waters do not or are not likely to meet the criterion for relevant classification under the Water Resources Act 1991 due to the land;
- any of the substances contained in Schedule 1 to the Regulations are the source of the pollution (this includes cyanide, mercury and mineral oil) of the waters;
- any part of the waters are contained within underground rock strata of the types specified in Schedule 1 of the Regulations.

Where any of these situations exist the Agency will take responsibility for the site but the initial determination will rest with the local authorities (subject to the right of the Agency to give notice to a local authority that they require a specific site to be designated under Section78C (4)). It is they who will inspect their area "from time to time" for the purpose of identifying contaminated land, and those sites falling within the categories specified in the 2000 Regulations will be designated as special sites and become the responsibility of the Agency. Local authorities must seek the advice of the Agency and have regard to such advice when making the determination. There is a mechanism for resolving disputes between the Agency and local authorities as to whether a site is "special", whether that determination comes from the Local Authority (Section 78C (1)) or the Agency (Section 78C (4)), with the referral of the decision to the Secretary of State for final determination. This appeal mechanism should be seen as a "last resort" as it would be unseemly for the enforcing authorities to be seen squabbling over their respective responsibilities.

In any event, attitudes towards acceptance of responsibility may well depend on how well individual local authorities are resourced both in terms of finance and expertise. This clearly points to the need for close liaison arrangements between the Area staff in the Agency and those individual authorities and the adoption of consistent approaches by each of them.

Remediation

In view of the difficulty of tracing persons who caused or knowingly permitted any land to become contaminated and the only requirement on the enforcing authorities being to make "reasonable inquiries" as to their whereabouts, it would have been helpful if the Guidance had given some indication as to how much effort and what steps the enforcing authorities should undertake in this exercise. The lack of such guidance will inevitably lead to appeals by owner/occupiers who receive remediation notices on the ground that the determination by the authority of them as the "appropriate person" is "unreasonable".

The other major problem area for the enforcing authority in respect of remediation is the fact that remediation notices cannot be served in respect of contaminated groundwater or surface waters where the person responsible is unknown or cannot after "reasonable inquiry" be found. In such a case, as noted in Chapter 5, Section 78J of the 1990 Act prohibits a notice being served on the innocent owner/occupier. Inserted, according to Government, to maintain consistency with the existing protection afforded to owners and occupiers in the Water Resources Act 1991; the consequence is that the Agency or local authority will have to undertake the remediation but will be unable to reclaim the costs of so doing under the contaminated land legislation. Clearly, the DETR expect the Agency to pick up this expenditure under Section 161 of the Water Resources Act 1991; the reality is, however, that in many instances the Agency will have to provide the funds for clean-up without any real prospect of recovery of its costs due to their inability to discover the identity of the person who caused or permitted the pollution.

The New "Clean-Up" Powers and their Interface with the 1990 Act Regime

Under Section 161 of the Water Resources Act 1991, the Environment Agency and its predecessor had powers to carry out either preventative works or, after a pollution incident, any necessary "clean-up" works and was able to recover the cost of such works from whoever caused or knowingly permitted the pollution or the potentially polluting situation. Since April 1999 the Environment Agency, by virtue of an amendment to Section 161 of the Water Resources Act, has been able to recover the expenses it has reasonably incurred in conducting an investigation to establish the source of the water pollutant and the identity of the polluter. Prior to this such costs were irrecoverable leading to unwillingness by the Agency to use this section.

The NRA had long campaigned for an additional power to require polluters to carry out clean-up works themselves and at their own cost, arguing that such a power would benefit the polluter as the costs it incurred in doing those works could be considerably less than those incurred by the Authority. As a consequence of this campaign Section 161A was inserted into the Water Resources Act (being effective from April 1999) giving the Environment Agency the power to serve a "works notice" requiring the person served to carry out pollution prevention works or, where pollution had occurred, to carry out clean-up works. Although it is clearly anticipated that the original power under Section 161 (namely for the Agency to carry out works itself) is to be used as very much a "last resort", the Agency is not precluded from taking its own action where there is an urgent need or the person responsible cannot be traced. Failure to comply with a "works notice" is in itself an offence, punishable with up to three months imprisonment and/or a fine of up to £20,000 on summary conviction or, on indictment, up to two years imprisonment and/or an unlimited fine. The Agency also has specific power to seek a High Court Order if it thinks prosecution for non-compliance would be ineffectual.

The works notice provisions potentially are a more flexible tool in the hands of the Environment Agency than the contaminated land provisions will be in the hands of local authorities. Under the new works notice provisions as long as the Agency considers a person caused or knowingly permitted the contaminant to be in a position where it could escape and cause pollution of controlled waters or caused or knowingly permitted pollution of controlled waters to actually occur, the power to serve a works notice is triggered. The Agency will not be required to, for example, undertake the type of risk assessment exercise that the local authority is required to undertake before classifying land as contaminated. Prior to serving a works notice however the Agency is obliged to consult with the person on whom it is proposed to serve the notice and there is a right of appeal against such notices. The required content of the works notice and the grounds of appeal are prescribed in the Anti-Pollution Works Regulations 1999.

One of the major concerns for the Environment Agency has been contaminated, dormant or derelict sites and the consequences of such sites in terms of pollution to the water environment, either groundwater or surface waters. Historically, the NRA investigated the water quality problem with a view to identifying the likely source of pollution and where the source could be identified, and the contaminated site was still occupied by the original polluter, the NRA's preferred approach, prosecution apart, was to seek to apply pressure on the polluter to undertake investigation and clean-up. Such an approach was undertaken in the knowledge that in the absence of such co-operation its only remedy was to exercise its own powers to clean-up under the then Section 161. However, cost recovery could only be directed against those who had generally caused or knowingly permitted the polluting matter to be in the "controlled waters" and even then depended on those persons being able to pay. Accordingly, there was always considerable doubt as to the suitability of the Section 161 powers for application to multi-million pound projects involving soil clean-up and removal. (Clean-up of polluted groundwater is an especially complex and lengthy process that is unlikely to be successful unless the problem of contamination is addressed at source, that is, the removal or treatment of the contaminated land). With the improved "clean-up powers" under Section 161 and 161A of the Water Resources Act and the introduction of the legislative framework for contaminated land, one can only speculate at this stage as to what policy the Environment Agency will adopt now in like situations.

However, in the short-term at least, the Agency is unlikely to depart substantially from the stance taken by the NRA, evidenced, for instance, in its highly-publicised Groundwater Protection Policy ("Policy and Practice for the Protection of Groundwater", the latest edition being 1998). The key feature of the policy is protection of groundwater through the inclusion of appropriate conditions in its own consents or via the planning process (when it is a statutory consultee). In addition there has been demonstrated a willingness (albeit somewhat infrequently) to prosecute under Section 85 of the Water Resources Act 1991 for pollution emanating from contaminated land sites and to exercise their clean-up powers under Section 161 of that Act.

As stated above, the Agency will have to determine how it uses its extended powers under Section 161A against the background of the contaminated land framework. Clearly, for the extended clean-up powers to operate, it will be necessary to establish who caused or knowingly permitted the pollution, whereas in relation to contaminated land, the requirement will be to find and serve a remediation notice on the appropriate person. However, it must be remembered that the Agency's jurisdiction in relation to contaminated land is in respect of "special sites" where the consequences of contamination are clearly of the most serious nature.

Against this background, it may be too simplistic to suggest that the Section 161A clean-up powers are not entirely appropriate for "special sites", but the fact is, as previously stated, there has always been considerable concern as to the suitability of Section 161-style powers for major site contamination with the prospect of the necessity for long-term remedial measures, and arguably this is still the case with the "works notice" powers. On the other hand, Section 161A does better reflect the "Regulator/Operator" principle and it may be that the Agency will therefore want to promote its use wherever possible.

It will be crucial for the Agency to address these issues and to produce a robust, unambiguous and sustainable policy that reflects fully the new legislative provisions and sets out how these new powers will interface in specific circumstances.

CHAPTER 9
DEALING WITH CONTAMINATED LAND IN COMMERCIAL TRANSACTIONS: SOME PRACTICAL ADVICE

The new contaminated land regime makes it more vital than ever when embarking on a transaction to take steps to identify the other party's environmental obligations and legal compliance record at an early stage in order to avoid the possibility of inheriting liabilities from the other party's failings. Buying land which turns out to be contaminated or buying a company with an environmental prosecution pending could prove very costly.

As detailed earlier, the new provisions require each local authority to cause its area to be inspected from time to time to seek out and ensure the clean-up of contaminated land, the passive management of land should now be positively discouraged.

It is increasingly common for banks and other funding institutions to insist on an environmental due diligence investigation before they commit funds. Sellers themselves are also beginning to carry out their own "pre-divestiture" reviews to establish where liability lies, whether and how it can be transferred from one party to another with the transfer of land or assets and how both owners and occupiers can protect themselves.

Everyone will be concerned with the cost of clean-up and be looking at ways to mitigate such losses. Indemnities between buyers and sellers are seen as a possible route by which sellers can transfer the cost burden. They will not, however, be able to transfer the statutory liability regardless of their ability to transfer the cost burden.

Many landowners are already cleaning up their sites but not all companies are in a position to do this. Small manufacturers and owners/occupiers are less likely to have the finances to do this, in which case they may look for a buyer to remediate.

The ways of inheriting environmental liability in both criminal and civil law have been with us for some time. The acquisition of the shares of a company will transfer to the purchaser the legal personality of the company and with it any pending environmental liabilities to which the company may be subject, including pending criminal prosecutions. In civil law some types of environmental liability can pass whether it is a company which is acquired or merely its assets. An owner/occupier of property may be liable under the law of nuisance for any damage caused by a "state of affairs" existing on the property. Even if he did not originally create that "state of affairs", if he adopts the nuisance by using the property for his own purposes or merely "continues" the nuisance by failing to take steps to abate it once it comes to his attention he can become liable.

What is new are the increased risks associated with environmental legislation.

The Environmental Protection Act 1990 introduced a number of radical new environmental protection regimes, including integrated pollution control, local authority air pollution control and new controls of the waste management and disposal. With these new regimes came tougher penalties for infringement and the potential for Directors and other corporate officers and managers to face criminal proceedings personally if their actions or omissions resulted in their company breaching those laws.

The public is now much more aware of environmental issues and there has been a radical restructuring of the way in which the environment is policed. Another factor making the inheritance of environmental problems more risky are the increased rights of the public to have access to environmental information held by public bodies under the Environmental Information Regulations 1992.

Any transaction involving land involves some form of risk-taking and risk-distribution. Today's methods of identifying the degree of contamination and pollution may not be tomorrow's and tomorrow's standards may be higher, thus it is still not an exact science and it should be accepted that even the most thorough site investigation may not fully establish the contaminative nature of the land.

Consequently, everyone involved in the transaction will want to negotiate the complete acceptance of a risk, either by way of a price reduction or by way of an indirect responsibility through and under warranty or an indemnity. Insurance may also be available to cover certain known risks.

So how should those owning or occupying contaminated land or land which has yet to be identified as contaminated now behave?

Let us approach this by dividing those who have interests in land into two categories, buyers and sellers; and then further dividing this latter category into those who know that the land which they own and operate is "contaminated", as currently defined, and those who have little knowledge of the condition of their land since they did not make any investigation regarding its state and condition at the time of their purchase.

The Buyer

These days there is no excuse for any properly advised buyer not to be aware of the risks he runs if he does not carry out a proper due diligence exercise at the time of his purchase. In any event, he may not have the option as his funder will invariably require him to discover as much as possible about the state and condition of the land he is buying prior to confirming any loan.

The maxim "caveat emptor" ("let the buyer beware") still applies, so the seller of land is under no duty to disclose any defects in the land. What the seller must not do is mislead the purchaser. The Government has sought to assist the purchaser with regard to this risk by trying to ensure that the purchaser is better informed by making available to him comprehensive information upon which he can take the decision whether to proceed or not. This is done via the 1992 Environmental Information Regulations which can be used by a buyer to ensure he gets from public authorities any information they have regarding the site's state and condition.

Any buyer should therefore try to obtain as much information as possible from the seller and, in the first instance, he will do so via his solicitor who will raise searching environmental preliminary enquiries before the buyer commits himself to the purchase. The buyer will want to know amongst other things:

- what activities are currently being and which have in the past been carried out on or at the site or any neighbouring or adjoining site;
- whether the seller knows of the existence of a watercourse within 250metres of the site;
- whether there are any or have ever been any underground storage tanks at the site or within 250metres of the site;
- whether the site has been infilled or the ground on any part of the site been raised;
- whether any regulatory body or other statutory body has served any notice or made any order in relation to the site or suggested that such notice might be served;
- the nature of the processes carried on at the site, the raw materials used, the by-products and waste products produced etc.

There is also no reason at this stage why a buyer should not commission "a desk top study", whereby through investigation of local and statutory authorities he can ascertain the various uses to which the site has been put in years past. The local authority may have a range of such information accumulated in anticipation of the requirement to maintain a register of contaminated land under Section 143 Environmental Protection Act 1990 prior to it being discarded. In addition, information will eventually be available by way of the register that must be maintained under Section 78R of the 1990 Act. Consequently the buyer should make comprehensive inquiries of the local authority, the Environment Agency and any other relevant statutory bodies.

Whether the buyer then decides to arrange for a site investigation and environmental assessment will depend on the value of the transaction and whether the costs of such investigations and assessments are commensurate with the value of such a transaction.

The buyer should also not forget to make comprehensive inquiries about the use to which adjoining and adjacent land has been put since it would be a tragedy if, having established that the site which he wishes to purchase is clean, he is then obliged to allow an adjoining or adjacent owner to come onto his land for the purpose of remediating the contaminated site next door. This could happen even after a buyer has developed the site.

Whilst a search of the Section 78R register may reveal the existence of a Remediation Notice, Remediation Declaration or a Remediation Statement, it will not reveal whether voluntary action is being or has been discussed with the particular landowner or what possible regulatory action the Local Authority is contemplating; direct inquiry about such actions should be made to the seller.

Buyers should also be made aware that even if a local authority has not identified the land they are buying today as contaminated, there is nothing to prevent them making such an identification at a later date as technology changes or the local authority receives more funding to enable it to be more diligent in its investigations.

Whilst the Act suggests that local authorities will have to be proactive, it is noteworthy that no additional funding is to be provided to them to enable them to undertake their investigatory responsibilities, so it may be that they will in fact only be reactive, acting on complaints. This reactive rather than proactive response is already evident in the statutory nuisance regime under Part II of the 1990 Act, which likewise requires the local authority to inspect it's area to discover the presence of statutory nuisances but which has developed into a complaint led system.

Once a site has been identified as a special site or as contaminated land then Section 78E (1) of the 1990 Act provides that the enforcing authority must serve "an appropriate person" with a Remediation Notice. Whilst the service of such a Notice is mandatory it is, however, subject to a consultation stage and there are constraints at the time of service and an embargo on service if certain conditions apply.

If after consultation it is agreed that, for example, voluntary action will be taken by the "appropriate person" then it can be agreed that the Remediation Notice need not be served and a Remediation Declaration or Remediation Statement will apply.

Consequently this consultation period may delay the service of a Remediation Notice for some time.

If the buyer wishes to obtain planning permission for a particular site before he commits himself to buy it he will need to ensure that any contractual obligation to obtain planning permission does not commit him to carry out an expensive site investigation which would outweigh the value of the site being purchased. Failure to draft an appropriate provision may obligate the purchaser to proceed on the ground of outline planning consent even though the outline planning consent obligates the purchaser to carry out an expensive site investigation. Hence, the planning process seeks to control the development of contaminated land through the development plan system, the development control system and through the use of conditions. However, it is an accepted principle that planning conditions should not seek to replicate environmental controls and thus any planning condition should be carefully considered.

Where the buyer's investigation reveals a degree of contamination it may be acceptable to both the buyer and the funder to continue with the purchase either on the basis of a reduction in the purchase price, which will reflect the cost to the buyer of remediating the site, or the seller may agree to remediate the property to a standard satisfactory to the buyer, the funder and the regulatory authorities. In this instance, the buyer will need to be satisfied that the seller is capable of procuring such satisfactory remediation and that the warranties or guarantees to be provided by the seller or the sellers contractor are worthwhile in that both are financially secure enough to ensure their existence for some time into the future when the buyer may have to call upon such warranties and guarantees.

A buyer experienced in purchasing contaminated sites may be content to carry out remediation himself provided the price and value of the transaction makes such a course of action worthwhile. In those circumstances, if the seller believes that the buyer is acquiring the site at a very attractive price and in full knowledge of the various contaminative uses which have been made of the site and of its liabilities for such uses, then the seller may seek to obtain an indemnity from the buyer to indemnify him (the seller) against all actions costs claims demands etc made against him as a result of such uses.

Whilst the seller cannot be released from his statutory liabilities as "the original polluter" or one who "caused" or "knowingly permitted" such pollution, he can seek to pass on whatever monetary penalty he suffers to the buyer who provides such an indemnity.

In such circumstances the buyer needs to be very carefully advised as to whether to provide such an indemnity.

Once the buyer has completed his due diligence investigation he may still consider it prudent to investigate whether insurance is available for any risks that he considers he should not take. Insurers will invariably specify extensive pre-conditions and, as is usual, the greater the risk to the insurer the greater the premium to the insured.

The Seller

The Seller who Knows That He is Selling "Contaminated Land"

In such circumstances various options are open to such a seller. He will be aware of his statutory liabilities and will know that he will not be able to avoid them if he is the original polluter or one who has caused or knowingly permitted such contamination. He may seek an indemnity from the purchaser whereby the purchaser will indemnify the seller against any clean-up costs and all claims demands etc made against the seller as a result of his or his predecessor's activities.

Alternatively, a seller who is currently or historically involved in contaminative uses may remediate the land himself or herself to a standard that should satisfy the more diligent investigation. Unfortunately, the absence of an industry norm for such remediation creates a degree of uncertainty, but nevertheless where companies such as Corus carry out such remediation buyers may draw a degree of comfort that such responsible companies they are seeking to make good the dereliction of the past.

The sellers should adopt responsible contaminated land management by:

- identifying the risks;
- acquiring/improving site knowledge and discovering if his land is contaminated;
- identifying and agreeing a management strategy to contain contamination and prevent future contamination;
- monitoring and/or remediating;
- not knowingly permitting by taking action to address known problems;
- avoiding remediation notices by taking voluntary action, consulting with the regulatory authorities and validating remedial actions;
- informing prospective purchasers of contamination and addressing liabilities with contractual arrangements to apportion responsibilities;
- liaising with the local community where there may be concern about the seller's activities;
- developing and retaining archive records and reviewing long-term monitoring requirements.

Sellers need to ensure that during the sale process they carefully control any investigation required by the buyer prior to the buyer committing himself to a purchase; a buyer may require a ground investigation and environmental assessment prior to exchange of contracts or may require exchange to be conditional upon satisfactory results of such ground investigation and environmental assessment. The seller should satisfy himself that the prospective buyer and his contractors retain full responsibility for any damage caused to the site as a result of such investigation and also for any exacerbation of any contamination either to the site itself or to any adjoining or neighbouring sites. The careful drafting of licences to allow access for such purposes is critical.

The Innocent Seller

Should all sellers of land carry out some form of pre-divestiture review so as to avoid being disadvantaged by a purchasers finding which might weaken their negotiating position, or should they permit the buyer to do his own due diligence investigation at his own cost? The answer to this question will depend upon the value of the transaction and the negotiating strength of each party. It may be that when the seller bought the property people were not aware of environmental issues and he may not have carried out even the most basic due diligence investigation. Such a seller may, in innocence, sell the property "in its present state and condition" and on the basis of the maxim "let the buyer beware", the seller may avoid all liabilities. But if the seller has been involved, to whatever degree, in any contaminative use or has caused or knowingly permitted any such use then he should seek to obtain an indemnity from the buyer for any action or demands made of the seller as a result of such use.

Leaseholders

Most leases have for many years sought always to ensure that tenants are responsible and liable for everything to do with the property during the currency of the term of the lease, thus having the effect of making the tenant liable for the costs of remedial measures in the absence of a landlord's repairing covenant.

If potential environmental liabilities between the landlord and the tenant are not catered for in leases then this will invariably give rise to problems in the future. Leases should now apportion responsibilities between a landlord and his tenant. Any prudent tenant should seek to exclude all liability for any breaches of environmental law arising from events occurring prior to the currency of his lease and require the landlord to provide an indemnity for any liability arising from such events. If the landlord is to remain liable for certain events then he must reserve a right of entry onto the land to enable him to carry out such remedial work. Leases should also have regard to the way in which the regulatory bodies discharge their duties and rights should be reserved to take account of this. Tenants may require leases to contain break clauses and rent suspension clauses should contamination prevent them from using and enjoying the site.

The apportionment of environmental liability may not in itself be sufficient and landlords would be very wise to ensure that tenants follow a prescribed scheme of environmental management throughout the term of their lease; failure to observe such a scheme may entitle the landlord to serve a Notice under Section 146 of the Law of Property Act 1925 and/or give the landlord the right to re-enter the property to remedy any contamination or pollution caused by the tenant.

"Pie-crust" leases are becoming more prevalent. The purpose of such a lease is to leave the contaminated soil out of the premises let to the tenant so that the tenant does not become the owner of the contamination "in, on or under land".

Warranties and Indemnities

Whilst acknowledging the limitations inherent in the taking of warranties and indemnities as a traditional method of protecting a party to a transaction against risk, they will continue to have a role in providing some protection against environmental risks. However sole reliance on them has a number of disadvantages:

- They are only as good as the credit-worthiness of the person who gives them. The ability to sue someone for breach of warranty is of little use if that party has gone into liquidation or has insufficient assets to cover the loss;
- In cases of pollution it is often very difficult to establish when it occurred and this may lead to an unresolved argument as to whether it occurred before or after completion and whether it is covered by a warranty which stated that the site was clean;
- Warranties may not have sufficient life span for them to be effective against environmental liabilities where contamination may only be discovered very many years after completion of the sale and long after warranties cease to be valid.

Specific warranties should be tailored to the environmental situation arising as a result of the due diligence investigation. The seller may be asked to warrant that he has obtained all necessary statutory permissions for his operations, that they are current and are transferable. They should also provide that regulatory bodies have not and are not taking any enforcement action under any statutory provision and that there are no pending civil actions being brought against the seller for environmental damage and that the seller is not aware of any circumstances which would justify the bringing of any such action. Warranties are commonly encountered in corporate transactions.

Warranties relating to the state of the land or equipment are often sought to confirm that there have been no spillages during the sellers occupation, but invariably sellers replies such as "only so far as the vendor is aware" limit the protection afforded by such a warranty.

Indemnities are a form of undertaking that the person providing such indemnities in specified circumstances will reimburse costs, claims demands etc incurred by the seller or buyer, as the case may be. In the absence of an indemnity the loss will lie where it falls and the giving of such indemnities will be determined by commercial considerations, such as the price being paid, the knowledge of each of the parties about the condition of the property.